THE ONTOLOGY AND MODELLING OF
REAL ESTATE TRANSACTIONS

T0231588

The Ontology and Modelling of Real Estate Transactions

Edited by

HEINER STUCKENSCHMIDT
Vrije Universiteit, Amsterdam, The Netherlands

ERIK STUBKJÆR
Aalborg Universitet, Denmark

CHRISTOPH SCHLIEDER
Universität Bamberg, Germany

Routledge
Taylor & Francis Group

LONDON AND NEW YORK

First published 2003 by Ashgate Publishing

Published 2017 by Routledge
2 Park Square, Milton Park, Abingdon, Oxfordshire OX14 4RN
711 Third Avenue, New York, NY 10017, USA

First issued in paperback 2017

Routledge is an imprint of the Taylor & Francis Group, an informa business

British Library Cataloguing in Publication Data
The ontology and modelling of real estate transactions. -
 (International land management series)
 1.Real property 2.House buying - Computer simulation
 3.House selling - Computer simulation
 I.Stuckenschmidt, Heiner II.Stubkjær, Erik III.Schlieder,
 Christoph
 333.3'3

Library of Congress Cataloging-in-Publication Data
The ontology and modelling of real estate transactions : European jurisdictions / edited by
 Heiner Stuckenschmidt, Erik Stubkjær, and Christoph Schlieder.
    p. cm. – (International land management series)
 Includes bibliographical references and index.
 ISBN 0-7546-3287-3
    1.Real property–Europe–Philosophy. 2.Conveyancing–Europe–Philosophy. 3.
 Ontology. I. Stuckenschmidt, Heiner. II.Stubkjær, Erik. III.Schlieder, Christoph. IV.
 Series.

 KJC1315.O58 2003
 346.404'37–dc21

                                                        2002043896

 ISBN 13: 978-1-138-27793-9 (pbk)
 ISBN 13: 978-0-7546-3287-0 (hbk)

Contents

PART IV: SYSTEMS ENGINEERING

List of Figures

List of Tables

List of Contributors

Armands Auzins is assistant at the Professor Group of Geodesy and Cartography, Riga Technical University, Latvia. His research interest include: Institutional analysis, cadastral and conveyancing procedures, cost assessment; terminology resources (concepts and terms). Contact him at: Professor Group of Geodesy and Cartography, Riga technical University Azenes street 16, room 109, Riga, LV-1048, Latvia phone/fax: +371 7089263, phone: +371 9439004 e-mail: armands.auzins@riga.vzd.gov.lv

Mark Deakin is Senior Lecturer and Teaching Fellow at the Napier University Edinburgh. His research interests include: real estate transactions, cadastral development, ICT applications, environmental assessment methodologies and sustainable urban development. Contact him at: School of the Built Environment, Napier University, 10 Colinton Road Edinburgh Scotland United Kingdom EH10 5DT

Robert Dixon-Gough is a member of the Land Management Research Unit of the School of Computing and Technology of the University of East London (UEL). His main research interest is in a comparative evaluation of Land Management Practices, with particular respect to Central and Eastern Europe. Contact him at: Land Management Research Unit University of East London Longbridge Road, Dagenham Essex, UK, RM8 2AS; Tel: (Home)+44 (0)118 971 3506 (UEL) +44 (0)208 223 2514

Hans Mattsson is Professor in Real Estate Planning (fastighetsteknik) at Kungliga Tekniska Högskolan (Royal Institute of Technology), Sweden. His research interests include Land Management and Administration. Contact him at: Real Estate Planning and Land Law Royal Institute of Technology S-100 44 Stockholm Sweden Phone: +46 8 790 86 17 Fax: +46 8 790 73 67 e-mail: mattsson@infra.kth.se

Chris Partridge works at the BORO Centre, London, England. His research interests include: Enterprise and top ontologies. Recovering ontologies from enterprise applications. Implementing ontologies in enterprise applications. Contact him at the BORO Centre; url: http://www.boroprogram.org

Christoph Schlieder is Professor for Applied Informatics in the Cultural and Geo-Sciences at Bamberg University, Germany. His research interests include: knowledge-based systems, spatial and temporal reasoning, applications of semantic information technologies to geoinformation systems and mobile computing. Contact him at: Faculty of Business and Applied Informatics, Bamberg

University, Feldkirchenstr. 21, 96045 Bamberg, Germany Phone +49 951 863 2840; Email Christoph.schlieder@wiai.uni-bamberg.de

Barry Smith is Julian Park Professor of Philosophy, University at Buffalo and Director of the Institute for Formal Ontology and Medical Information Science, University at Leipzig. His research interests include: Formal and applied ontology, especially medical ontology; applications of ontology to spatial information science. Contact him at: Department of Philosophy, University at Buffalo 130 Park Hall, Buffalo NY 14260, USA; phone: +1 716 645 2444; fax: +1 419 781 8794; email: phismith@buffalo.edu; url: ontologist.com

Milena Stefanova is researcher at CNR in Padova, Italy. Her research interests include: business process modelling and analysis, billing and payment methods and technology, security and trust in e-business. Contact her at: CNR, Istituto ISIB, Corso Stati Uniti 4, I-35127 Padova, Italy; email: milena_stf@yahoo.com

Erik Stubkjær is Professor for Cadastral Science at the Department of Development and Planning, Aalborg University, Denmark. His research Interests include: Contribution of scientific enquiry to cadastral development, primarily in Europe; ontologies and semantic web; web technology and its application in university education; organisational aspects of geographical information systems. Contact him at: Department of Development and Planning, Aalborg University, Fibigerstraede 11, DK-9220 Aalborg East, Denmark, phone +45 96 358350; fax: + 45 98 156541; email: est@i4.auc.dk; url: www.i4.auc.dk/est/

Heiner Stuckenschmidt is post-doc researcher in the Artificial Intelligence Department of the Vrije Universiteit Amsterdam. His research interest include: ontology-based information modelling and integration, knowledge representation and reasoning on the semantic web. Contact him at: AI Department, Vrije Universiteit Amsterdam, de Boelelaan 1081a, 1081 HV Amsterdam. phone: +31 20 4447752, fax.: +31 20 4447653, email: heiner@cs.vu.nl; url: http://www.cs.vu.nl/~heiner/

Radoš Šumrada is Professor for GIS technology and science at the Department of Geodesy, Faculty of Civil and Geodetic Engineering, University of Ljubljana, Slovenia His research interests include: Contribution of scientific enquiry to GIS technology domain and cadastral development in Europe; geographical information systems and its application in university education. Contact him at: Faculty of Civil and Geodetic Engineering Jamova 2, SI 1000 Ljubljana, Slovenia; phone +386 1 4768650; fax +386 1 4250704; email: RSumrada@Fgg.Uni-Lj.SI; url: http://fgg.uni-lj.si/home.htm

York Sure is PhD candidate at the Institute AIFB, University of Karlsruhe. His reserach interests include: Semantic Web, methodologies for developing and employing ontology based knowledge management applications. Contact him at: Institute AIFB, University of Karlsruhe D-76128 Karlsruhe, Germany;

phone: +49 721 608 6592; Fax +49 721 608 6580 email sure@aifb.uni-karlsruhe.de; url: http://www.aifb.uni-karlsruhe.de/WBS/ysu

Harry Uitermark is Geographic Information Systems consultant. Contact him at: The Dutch Cadastre P.O. 9046 7300 GH Apeldoorn, The Netherlands email: harry.uitermark@kadaster.nl; phone: +31 (0)55 5285506 www: http://www.kadaster.nl

Kauko Viitanen is Professor for Real Estate Economics and Valuation; Dean Institute of Real Estate Studies, Department of Surveying Helsinki University of Technology, Finland. Helsinki University of Technology P.O. Box 1200 FIN-02015 HUT, Finland phone +358 9 451 3870; fax + 358 465 077; email kauko.viitanen@hut.fi ; url: http://www.hut.fi

Ubbo Visser is Assistant Professor working in the Artificial Intelligence group at the Center for Computing Technologies (TZI) within the Department of Mathematics and Computer Science at the University of Bremen. Research interests include: Knowledge representation and processing for the Semantic Web, Multiagent systems in dynamic and real time environments, Environmental Applications of AI systems and Hybrid integrated stems. Contact him at: phone: +49 421 218 7840; fax: +49 421 218 7196 email: visser@tzi.de

Leo Zaibert is Assistant Professor of Philosophy at the University of Wisconsin-Parkside. His areas of interest include: Philosophy of Law, Philosophy of Mind, Political Philosophy. Contact him at University of Wisconsin-Parkside; phone: +1 262 595 32 36; email: Zaibert@uwp.edu

Foreword

Successful management depends on both, knowledge about the domain under consideration and information about the current state of affairs. This holds for land management as well as for any other management activity. For the special case of land management the knowledge and the information involved has to do with land and its use, its development over time and many other properties. In this book, we focus on the specific problem of capturing and exchanging knowledge and information about the specific aspect of property rights to land at a European level.

The significant differences between the legal systems in Europe make conventional comparison approaches hard to apply. The work reported in this book therefore chooses a new way by applying modelling techniques from computer sciences, in especially from software engineering and artifical intelligence. Both disciplines have a long tradition in the development and application of methods for describing knowledge and information in a comprehensable way.

This book is organized in three main parts that cover the overall problem statement as well as a decription of technical tools that could help to approach the problem of modelling the different systems and processes in order to achieve conceptual and terminological clarity.

Part I introduces the domain of real property transactions on a gerenal level. The authors contributing to this part decsribe the domain from different perspectives providing the reader with an impression of the complexity and the problems that occur when trying to capture knowledge and information. Stubkjær gives an overview over goals, problems and research questions connected with modelling real property transactions from a multidisciplinary perspective by introducing the European COST G9 'Modelling Real Property Transactions'. Mattsson elaborates the legal aspects of the problem domain by emphasizing the role and the importance of property rights with repsect to land markets and the economy in general. Zaibert and Smith raise some fundamental ontological problems of real property transactions from a more philosophical perspective and identify some core concepts connected with real estate and ownership that have to be clarified in order to reach a common understanding of the domain.

Part II leaves the general level of investigation. The contributions to this part decsribe the actual system for handling real property transactions in selected European contries. The regulations and processes described illustrate the diversity of organizational and legal structures we find at the European level. Beside their descriptive character, the contributions to the second part of the book provide case studies that can be used to identify requirements of a general approach for modelling real property trasactions and to validate available technology and proposed solutions. Viitanen describes the Finish system by choosing a particular transaction type, namely the purchase of a house and gives a detailed description of the necessary steps as well as the decisions involved. Dixon-Gough and Deakin give an overview of the situation in the United Kingdom and motivate the existing differences between England, Scotland and Wales by a of the history of the different systems. Auzins, finally, describes different types of real property transactions in Latvia and relates them to the societal background of the country.

Part III already describes parts of the approach we envision for enabling people to exchange knowledge and Infomartion about real property, property rights and exchange of these rights. This part discusses ontologies as an enabling technology for explicating knowledge about the domain under consideration. In contrast to the philosophical view of Zaibert and Smith, the notion of ontology used in this part of the boof is a more pragmatic one as it borrows from knowledge and information modelling rather than philosophy. Visser and Schlieder disucss the role ontologies can play in achieving the goals described in the first part of the book. They give a definition of ontologies and review modelling techniques for ontologies as well as attempts to build ontologies for domains related to the topic of the book, i.e. legal ontologies and ontologies of space and time. Sure proposes a methodology for actually building ontologies about the subject area. The author decsribes necessary steps and introduces technology that supports ontology development especially in the early stages of development. The contribution of Partridge and Stefanova is concerned with experiences that have been made with building ontologies about related areas, in this case with modelling organizational structures. They show the difficulties in coming up with a universally valid definition of terms by comparing different organizational ontologies.

Part IV substantiates the claim that modelling techniques known from computer science and especially ontologies can indeed be used to capture knowledge and exchange information about real property and cadastral systems. The two contributions to this part describe successful applications of conceptual modelling techniques to particular problems related to real property transactions. Sumrada illustrates the use of the Unified Modelling Language UML in the development of cadastral information systems and argues that the development process can be significantly improved using conceptual modelling. Uitermark presents the use of ontological modelling to achieve interoperability between geographical datasets which is a prerequisite for exchanging information about real property transactions.

Summarizing, the book covers the whole range of topics connected to the modelling of real property transactions from principled problems, specific situations in different European problems via technology for supporting the modelling process to some partial solutions to the problem that make us optimistic about the possibilty of coming up with more complete solutions to the problem of capturing, exchanging and comparing knowledge and information about real property transactions on a European level.

Heiner Stuckenschmidt
Erik Stubkjær
Christoph Schlieder

PART I
CADASTRE, LAW AND ECONOMICS

Chapter 1

Modelling Real Property Transactions

Erik Stubkjær

Abstract

The chapter introduces the COST research action *Modelling Real Property Transactions* by presenting its societal and scientific context, and developing on the research challenge. The primary societal context is the Commission of the European Union. The scientific context presented includes the research fields of land management, knowledge engineering, and law and economics, respectively.

The research action is described with reference to New Institutional Economics. It aims at the specification of institutional functions within the domain of real property. The specifications draw upon and contribute to research in legal and geographical ontologies, and are assisted by computer tools under development. The institutional functions of the domain of real property are developed from the modelling of national, legal procedures by means of Unified Modelling Language, or similar formalisms, and consequtive comparative analyses of the models across selected European countries. Transaction costs are assessed on the basis of the modelled transactions.

Introduction

Families buy houses to make homes. Farmers and companies buy land and other real estate to extend their business. Municipalities acquire land to provide for roads and other technical infrastructure. These activities, in technical terms: *transactions* constitute a transfer of real property rights. The transactions are accomplished by the signing of documents, which are circulated in a complex configuration of professionals and public bodies. Professions include lawyers, notaries, geodetic surveyors, and real estate valuators, who support the owners, as well as authorities, with their expertise.

Property rights are protected by national constitutions and mentioned in the UN Declaration of Human Rights, art. 17. Yet, property rights can be maintained and enforced only through governmental support. In European countries, the transfer of real property rights is mostly recorded at the land registry section of the courts. The identification of individual real estates is mostly achieved by cadastral authorities, which maintain information systems with property records and maps. The transfer of property rights includes the conveyance of title and mortgaging. The transfer processes are closely related to changes of the extension of the property, and to the formation of new parcel lots. The transactions and the intended future use of real estate have to comply with spatial planning, and other agricultural and environmental legislation. Finally, the processes and the stock of real estates are used for the collection of a variety of fees and taxes and for the provision of statistics.

The terms *transaction* and *transaction costs* are technical terms within economics. They relate to the fact that the cost of a commodity in a market reflects not only the price paid. The cost includes the efforts of searching for the relevant commodity and of assessing the quality of the product, as well as the costs of legal protection of property rights, including enforcement measures (North, 1996). The costs of transfer of real property rights depend, among others, on the efficiency of public administration. It is an open question to which degree the systems and the processes can be privatised with economic benefit.

There is good reason to address real property transactions and their costs through a joint European research project. For example, in some European countries notaries must be involved in property transactions, while in other countries this is not the case. Similarly, subdivision of land is tightly regulated in some countries, but this is not generally the case. These examples suggest that the potential for improving efficiency of transactions and of market operations be far from exploited. This is likely due to the fact that the transaction processes are complex and difficult to delineate from other societal activities. The transactions are regulated by rules, but the observance of rules may fluctuate as the norms and cultures permit.

In economic terms, the field is noticeable. A study by the Economist Intelligence Unit, London, suggest that the real estate component of foreign direct investment in the world economy '..could be anywhere between 5-20 per cent of the total. Either way the real estate sector is significant..'. One of the conclusions of the study was that ' .. Global harmonisation of property markets and standardization of rules and regulations governing real estate are necessary steps to boosting investor confidence and allowing transparency in the FDI regime governing real estate' (EIU, 1997, as quoted by Dale and McLaughlin, 1999: 5).

Studies of transaction costs have been made. Table 1 summarises the findings of a desk study, commissioned by the UK Department of the Environment, Transport and the Regions to Barony Consulting. The figures, here in integers, do not include equivalents of stamp duty and VAT.

Table 1: Home buying costs as per cent of £50,000 property (before tax)

Country	Costs (%)
Australia	3
England and Wales	3
Sweden	5
Denmark	6
France	6
United States	6
South Africa	8
Portugal	10

Source: Office of the UK Deputy Prime Minister. Housing Market Transactions: International Comparisons (*Housing Research Summaries* No. 77, 1998) Abriged.

Transaction costs concerning the refinancing of residential property have been compared. The outcome shows that it takes considerable longer time to process such a transaction in the United States, and at a considerable higher cost than the most advanced European jurisdictions. For the US the figure 6.546 USD is thus compared to 1.035 USD for Sweden (Kjellson, 2002).

The above findings support the need for a better understanding of this area. In order to establish such proven understanding it is appropriate to restrict the field of study to selected European countries. Even a neighbouring pair of countries, e.g. Denmark and Sweden, or Austria and Slovenia, has remarkable differences, which makes it a challenge to elicit a common set of concepts and models.

COST, European Co-operation in the Field of Scientific and Technical Research, supports the co-ordination and networking of existing research activities, but it does not fund research itself. Rather, COST funding covers the co-ordination expenses of each action (scientific secretariat, contribution to workshops and conferences, publications, short-term scientific missions etc).

A joint European proposal for the project *Modelling Real Property Transactions* was adopted as COST action G9 as of March 2001, when representatives of Austria, Denmark, Germany, Netherlands, and Spain signed the Memorandum of Understanding (COST MoU 328/00). By the end of the year, the number of participating countries increased to 11, including the non-EU countries Hungary, Latvia and Slovenia.

The implementation of a COST Action is supervised and co-ordinated by a *Management Committee*, who elects a chairperson, the present author. The following addresses the project as a whole, but does not necessarily reflect the opinions of the Management Committee.

The Memorandum of Understanding includes a Technical Annex, which describes the foreseen research. The Technical Annex was prepared late in 1999 and - with only a few, minor editions - serves as terms of reference for the COST action. The following is based on the Technical Annex, but reworked to provide an updated account of the research issue. The following sections present the background (section 2) and objectives and benefit (3) of the scientific programme, and develop on state of the art (4), the research issue (5), and methodological considerations (6). A conclusion closes the chapter.

Background

The scientific programme of the COST *action* G9 is performed in a specific societal context, which is different from the research context, described the subsequent section 4 on state-of-the-art. The following description of the societal context draws largely upon the conception of society, which is expressed through the organisational structure of the Commission of the European Union.

The internal market

The *internal market* of the European Union has facilitated citizens and companies, who want to purchase real estate in another EU country. The purchasing procedure is complex in any country, and hence presupposes professional assistance. Consequently, cross-border professional service is getting increased focus, as illustrated by the following three examples.

The European Commission (EC) is monitoring the implementation of Community law and addressing infringements, e.g. regarding professional services. The EC decided on 3 July 1998 to institute proceedings in the Court of Justice, following the Spanish Government's reply to a reasoned opinion notified on 27 January 1998. This is because of the obligation for non-Spanish nationals wishing to buy real estate to use the services of a Spanish notary, while there is no such obligation for buyers resident in Spain (EC, 1998). However, the case was closed before executing the decision to seize the Court of Justice, because Spain cancelled the condition of taking a Spanish notary contained in the Royal Decree 671/92 by the Royal Decree 664/99 of 23.4.99 (OJ, 1999).

Interestingly, in the same period notaries introduced common standards of service on a European level. *The European Code of the Notarial Professional Ethics* (Deontology) was adopted in 1995 by the Conference of the Notarial Associations of the European Union in Naples, and modified in 2000. It sets out ethical principles such as independence, confidentiality, impartiality and also conditions of a notary's function such as training or professional indemnity insurance. Important in the present context is the rule that always the local notary must be in charge: 'Jedenfalls darf nur der territorial zuständige Notar beurkunden.' (Bundesnotarkammer, 2000:2.1).

Thirdly, the European Mortgage Federation, which represents European credit sector associations, and European consumer organisations in March 2001 signed the *European Agreement on a Voluntary Code of Conduct for Pre-contractual Information on Home Loans*. By endorsing the Code, the lending institution agrees to offer specific information. A summary of the loan conditions will be presented in a standardised format, a European Standardised Information Sheet (ESIS). The standardised documentation of offers for home loans will help prospective borrowers to choose their mortgage loan and compare products both on a national and cross-border basis. This will help develop cross-border mortgage transactions and further competition in the European mortgage market. The European Commission records the mortgage lending institutions, who are prepared to sign the Code of Conduct, and provide further information on the scheme and its implementation. The signing mortgage lenders are to implement the scheme by September 2002 (European Mortgage Federation, 2001).

Other initiatives with respect to the internal market and property transactions in Europe include the Commission's action plan for skills and mobility. The plan notes that regional wage differentiation, the design of the tax-benefit system and the functioning of the housing market strongly affect the propensity to move. 'A crucial element affecting the housing market is the extent to which there are low transaction costs, such as taxes and fees for real estate agents, notaries and land registration' (COM 2002 72: 10). Moreover, the surveying profession of the EU countries in 1988 became represented at the EU Commission through the CLGE, *Comité de Liaison des Géomètre-Experts Européens*. The CLGE facilitates the mutual recognition of professional qualifications, among others by describing the education and practice of the geodetic surveyor in Western Europe. Furthermore, heads of the mapping agencies in Europe have organised *EuroGeographics*, formerly CERCO. EuroGeographics is a voluntary association of the heads of European mapping agencies. It encourages collaboration and the exchange of information on the matters of mutual concern between members and also assists in the creation of the European Spatial Data Infrastructure. As several mapping agencies are not concerned with land management issues, the agencies with cadastral and further land administration activities have organised a *Working Party on Land Administration* within the UN-ECE. The WPLA aims at promoting land administration through security of tenure, establishment of real estate markets in countries in transition, and modernisation of land registration systems in the market economies.

The Information Society

The above-mentioned creation of codes for notaries and for the marketing of home loans, constitute an alternative to a traditional top-down regulatory approach. Rather, it is in line with the open method of co-ordination, applied since 1997 for the European Employment Strategy (cf. Goetschy, 1999). Lisbon European Council in March 2000 gave the open method increased emphasis and wider scope. In line with this trend, the communication on *the exploitation of public sector information* also refers to the open method, e.g. by supporting exemplary projects and stimulating the exchange of best practices throughout Europe (COM 2001 607: 10).

The development of an Information-based Society was seen as the key to the development of new job opportunities in the Commissions White Paper on Growth,

Competitiveness, and Employment published in December 1993. The subsequent Bangemann-report triggered the process of liberalisation of the telecommunication sector in Europe during the 1990s. Next step was to support the availability of information content, fit for the established tele-infrastructure. The Green Paper on Public Sector Information: a Key Resource for Europe (COM 1998 585) suggested making public sector information in digital format available for its reuse beyond the purposes for which it was originally collected.

The recent communication on exploitation of public sector information mentions 'Legal information (in particular concerning legislation and jurisprudence) ..' and 'Geographical information relevant to transport and tourism ..' (COM 2001 607: 3). The communication considers public sector information a prime content resource and an important asset with a substantial growth potential and foresees an increasing demand for pan-European information products. The United States, where citizens and companies enjoy a broad right to electronically access public information and reuse it for commercial purposes, is seen as the model to imitate.

Table 2: Geographical data groups and their relative economic weight

Topographic objects: 33%		**Properties, etc: 29%**	
Transport	5	Administrative units	2
Relief and contours	7	Ownership units	27
Hydrography	5	Addresses	?
Other environmental	16		
Utilities	19		
Geodetic network, etc.	4		
Maritime navigation	15		

Source: ETeMII, 2001: 9ff and Annex C: 36, data from ANZLIC, 1995. Edited

A specification of legal and geographical data is available from other documents, provided by a project consortium (ETeMII, 2001), and drawing on figures from investigations on geographical information in Australia (ANZLIC, 1995). The figures of Table 2 may give a rough idea of the relative sizes of cost of production for original use. The sources do not reveal the relation of the total (100 per cent) to the turnover of the Australian mapping industry and, furthermore, do not indicate how geographical data (maps) for road and building construction are counted. The latter could amount to the same order of size as the Utilities.

According to the information available, data on Ownership units is, relatively, the largest group of data (27), followed by data on environment (Hydrography and Other environmental: 21). The concern for environmental data is reflected by the recent effort of DG Environment to develop an Environmental European Spatial

Data Infrastructure (EESDI) within the frame of the Water Framework Directive 2000/60/EC.

The communication on *exploitation of public sector information* notes that rules and practises for re-using data diverge between countries, or are not clear. Therefor, high quality information on administrative procedures and investment conditions is requested, among others. Proposed EU actions include the creation of pan-European meta-data and standards, as well as the provision of public sector information through portals at national and pan-European level. The exchange of best practises, and comparative case studies, establishing an *e*Europe framework, is one of four main actions suggested.

Enlargement, the rule of law, and real property

As mentioned in the introduction, real property rights are supported only if land registries and cadastral agencies are functioning and able professionals and civil servants are available. These and other components of an infrastructure of property rights, land markets and administration (cf. Holstein, 1996: 13) are part of national powers, and thus generally not within the scope of the Commission of the European Union. However, the process of enlargement of the European Union with Central and Eastern European countries made the Commission refer to this infrastructure in official documents, as we shall see. Furthermore, a study of land markets in Central and Eastern Europe was made under the Action for Co-operation in the field of Economics (ACE) programme of the European Union (Dale, Baldwin, 2000).

Briefly stated, the Copenhagen accession criteria of 1993 consists of three main requirements: stability of institutions guaranteeing democracy and the rule of law, the existence of a functioning market economy, and adherence to the aims of political, economic and monetary union. Furthermore, in Madrid the European Council urged candidate countries to adjust their administrative structures, and at Luxembourg, it stressed that incorporation of the acquis into legislation is necessary, but not sufficient; it is necessary also to ensure that legislation is actually applied. These general requirements, organised into 31 chapters, are negotiated in detail with each accession country.

Chapter 4 on *Free movement of capital* addresses one of the Copenhagen economic criteria, the existence of a functioning market economy. Among accession criteria is thus:

> The legal system, including the regulation of property rights, is in place; laws and contracts can be enforced;

The perspective is free movement of capital in terms of foreign direct investment. However, the investment perspective on real estate co-exists with other perspectives. For example, individual ownership is often heralded as a means to secure social stability and as an incentive for owners to invest their resources in their estate and thus obtain better living conditions and higher productivity of the estate. Furthermore, the satisfaction of being master of one's house need not be restricted to the micro level of individual units of real estate. Indicative in this respect is the provision established in the context of Danish entrance in the European Union: 'Notwithstanding the provisions of this Treaty, Denmark may maintain the existing legislation on the acquisition of second homes' (Maastricht Protocol, 1992).

Several candidate countries have thus requested, and been granted, transitional periods on foreigners' right to investment freely in real estate. The general EU proposal on purchase of real estate is

- a five year transitional period for the acquisition of secondary residences, excluding EEA citizens (citizens from the EU and Norway, Iceland and Liechtenstein) who reside in the future member state from the scope, and
- a seven year transitional period for the acquisition of agricultural and forestry land, excluding self employed farmers from the scope.

The statuses of fulfilment of the requirements are recorded in Commission Opinions (1997), in the Council Decisions on accession partnership (1998), in yearly Regular Reports (1998-), and recently in a survey of Accession Negotiations: State of Play January 2002. For example, the needed commitment for short-term economic reform in Latvia includes 'modernisation of the agriculture sector and establishment of a land and property register' and the item: Reinforcement of institutional and administrative capacity includes 'the establishment of a training strategy for the judiciary, ..'. For Slovenia, short-term commitments regarding Reinforcement of institutional and administrative capacity emphasises ' .. , improvements in the areas of the judiciary, of land registration, ..' (Council Decisions, 1998), while Council Decisions for Hungary did not mention provisions in this field. A composit paper notes that 'None of the candidate countries have demonstrated significant progress in the area of agricultural structural reform. ... All of the countries need to adopt a more comprehensive approach to aligning their policies and practice to those of the EC in areas such as market and price organisation, rural development, land structure and ownership etc.' (EC, 1999:28). Specifically for Romania and Bulgaria, the lack of a functioning land market is recorded (: 42, 43).

Summary

Closing this section, it can be noted that real property transactions and the related markets, agencies, professionals, and information systems are addressed by more than three of the Directorates-General of the Commission. Not covered here are, for example, projects that have been launched with European support in Greece and Portugal, with a view to improve their land registration and recording of title (cf. OJ, 2001).

Within the context of the Information Society and the exploitation of public information, available data suggest that information on property is the largest among geographical data groups, followed by data on environment, and on utilities, respectively. In COM (2001) 607, the Commission – as one of four main actions – calls for the exchange of best practises, and comparative case studies, establishing an eEurope framework. More specifically, a need is stated for the creation of pan-European meta-data and standards, as well as the provision of public sector information through portals at national and pan-European level.

Responding to the growing internal market, professions like geodetic surveyors and notaries have established European organisations to accommodate their services to the internal market. The notaries have introduced common standards of service in terms of a notarial professional ethics. Similarly applying the open method of co-ordination, mortgage lenders adopted a code of conduct for information on home loans.

Finally, in the context of the enlargement process, the Commission perceives the legal system, including land registry, and related functions, as prerequisite for a functioning market economy, in fact as a vehicle for foreign direct investment.

Objectives and benefits

Objectives

The main objective of the COST action G9 *Modelling Real Property Transactions* is to improve the transparency of real property markets and to provide a stronger basis for the reduction of costs of real property transactions by preparing a set of models of real property transactions, which is correct, formalised, and complete according to stated criteria, and then assessing the economic efficiency of these transactions.

The property transactions are stating or changing property rights and parcel lots. From a modelling perspective, the transactions are accomplished through inter-organisational business workflows. For selected European countries a comparative analysis will be made with a view to identify common functional units of the business workflows. A core set of such institutional functions will be rigorously described in English and the language of the selected countries. The detailed information will be presented in such way as to include a formal description of the underlying data. The models of real property transactions must satisfy the criteria of validity from an information modelling, ontological perspective, as well as from a legal perspective.

Similarly for selected countries, a comparative analysis of the economic efficiency of transactions involved in the transfer of property rights will be presented. The essential effects, intended and non-intended, of the real property transactions are likely to differ among the countries being investigated, due to institutional differences. The comparative analysis of the economic efficiency of transactions will include an identification of these effects and an assessment of their impact on the economic efficiency, including an assessment of the value of transaction information for further purposes. An exploratory analysis of relations between transaction costs and national practices regarding land management, education and governance may be added.

Benefits

The main benefit of the action is that governments, professions, and holders of property rights get a better basis for reducing the costs of the transactions of the markets of real estates. The developed models will increase the transparency of real property transactions. The present rules and notions of conveyancing of property rights may be carefully prepared, but from a comparative perspective are often odd. These rules and notions are rephrased into rationally structured information (ontologies). Information material can be derived from the ontologies. Provided that such material is produced and disseminated, e.g. through web portals, *owners* get a basis for being better prepared for professional advice, a situation already realised within medicine.

The developed models can inspire *professional organisations* of European scope to reflect the present diversity of procedures for solving similar tasks, and review and detail the more tentative outcomes of the COST action. For example, the outcome of the comparative analysis of the economic efficiency can be used as a base for improving the efficiency of the procedures. For *governments*, the models can be used for drafting new ordinances, and staff at *university departments* will use them for education.

Mortgage and other financial institutions, which are operating on a pan-European scale, may benefit from the comparative analyses. Following further research and development, pan-European networks of portals serving real property transactions can be envisaged, among others developing from the portal, which is going to hold information on house loan mortgage institutions. The information provided will extend the markets and make them more transparent. Information on property, services, and prices can be compared more easily, and thus enhance competition and efficiency.

The COST action G9 provides added value: the European accession countries, for shaping their institutions of real property urgently need The action's outcome in terms of models and analyses. The experience gained by the early participation of Hungary, Latvia, and Slovenia within the research network will assist other countries in their transition efforts. Furthermore, the COST action will support PhD-studies by providing a much-needed international research framework and a basis for PhD-level courses. Finally, the research may refocus the perspective on real property to its basic man-land-society context, and thus support European dialogue with other countries and cultures.

State of the art

The proposed network will draw upon recent research from the diverse views of land management, formalisation of information, law, and economic theory.

Land management

The term *land management* is a comprehensive expression for activities regarding land resources (cf. Larsson, 1997). The term land administration is more narrow and refers to 'the processes of recording and disseminating information about the ownership, value, and use of land and its associated resources' (ECE, 1996).

International concern for these matters is of a rather recent date. The 1970s saw an emerging interest for the 3rd world countries and their development, including the development of land management. The World Bank and UN bodies, e.g. the United Nations Centre for Human Settlements UNCHS (Habitat) thus provided a frame for research and development in land management issues. Around 1990, textbooks of an international scope were issued (Dale and McLaughlin, 1988; Larsson, 1991).

Both textbooks use the term Land Information as a general term. This was in accordance with the term Land Information Systems that was adopted in 1978 by the International Federation of Surveyors (FIG) for one of their scientific commissions. The textbook by Dale and McLaughlin present a taxonomy of information systems where Land Information Systems is the term applied for systems related to large map scales (generally used for cadastral purposes), while Geographical Information Systems is the term applied for small map scale systems (mostly used by geographers). The latter term became, however, the general term for the rapidly developing research field, cf. Longley et al. (1999). Regarding the legal issues, both textbooks address the registration of rights in real property, but it appears that the authors take an information system, rather than a legal approach. The two textbooks introduce economic and feasibility issues in terms of rational analysis of problems, assessment of benefits and costs, decision on Land Information project, implementation, and monitoring. The established system will in turn improve the decision-making. Also, both textbooks refer to a paper by G. Feder for a World Bank seminar in 1986. The reasoning goes that titled land provides security to farmers as well as to lenders, which will trigger more investment. The increased investment provides for more variable input use, which in turn gives higher output, higher income, and higher prices on land.

Recently, Hernando de Soto demonstrated that people in developing countries hold assets in terms of real property that 'far exceed the holdings of the government, the local stock exchange and foreign direct investments; [and that these assets] are many times greater than all the aid from advanced nations' (2000: 28). However, the poor cannot participate in an expanding market, because they do not have access to a legal property rights system that represents their assets in a manner that makes them widely transferable and fungible. Economic reformers have left the issue of property for the poor in the hands of conservative legal establishments, which are uninterested in changing the status quo. As a result, the assets of the majority of their citizens have remained dead capital stuck in the extralegal sector (2000: 192f). Other accounts of development projects are made as well, e.g. by Holstein (1996). Williamson and Fourie note that many cadastral reform projects do not respond to the expectations of those that conceived them, and call for better research methods. In adapting the case study methodology to cadastre, Williamson and Fourie proceed to present methods for data acquisition, which include interviews and participant observation. They build a 'Cadastral reform methodology' comprising three stages: case studies, comparisons and solutions (1998).

The complex process embedding property transactions is addressed from the point of view of benchmarking (Steudler, Williamson, Kaufmann and Grant, 1997), or with a view to chart the interrelated technical, legal, and organisational aspects (Zevenbergen, 1998). Stubkjær surveys research in information systems development and research within geographical information science with a view to establish a theoretical basis for cadastral studies (1999), and Silva and Stubkjær review methodologies used in research on cadastral development (2002).

Although the above research is encouraging and continuing, e.g. (Lemmen and Oosterom, 2001), it also demonstrates that presently there is still an insufficient understanding of the instruments (in a broad sense) which are necessary and sufficient for establishing and sustaining markets in real property rights.

Formalisation of information

An early example from the perspective of formalisation is 'An object-oriented, formal approach to the design of cadastral systems', presenting a simplified ontology of an integrated cadastral-land registry unit by means of a functional programming language (Frank, 1996). Later, in the paper 'The Structure of Reality in a Cadastre', Bittner, von Wolff and Frank apply Searle's theory of social reality in order to understand the institutional structure that the cadastre is intended to represent correctly. Searle's theory introduces an explanation on how an institutional reality is constructed on the base of a physical reality. It establishes that the institutional facts exist because people agree on them, there is an element of 'collective intentionality'. Institutions are defined as 'sets of constitutive rules' (2000: 88, 92). Ownership is an institution, and being an owner is an institutional fact.

Formal, applied and geographic ontologies have been researched at the Cognitive Science Center, University of Buffalo, USA. 'The Metaphysics of Real Estate' (Smith and Zaibert, 1997) develops upon the fact that real estate is a complex, historical product of interactions between human beings, legal and economic institutions and the physical environment, as it is used and occupied. The perspective includes the treatment of land property in industrialised nations as well as land allocation in tribal cultures. An analysis into the nature of property boundaries has provided an important cue on the interplay between visible terrain objects, and the humanly constructed, legal facts that generally are not visible (Smith, 1995).

In Europe, the Intelligent Systems Group at the Center for Computing Technology, University of Bremen, has developed 'Ontologies for Geographic Information Processing' (Visser et al, 2002).

Another perspectives on property transactions include that of business workflows, cf. work on enterprise ontologies that is reviewed in this volume (Patridge and Stefanova), and the perspective of spatial and temporal object databases. The latter line of research has been pursued by several European research projects, e.g.:

- 'DISGIS - Distributed Geographical Information Systems. Models, Methods, Tools, and Frameworks' ESPRIT Project Nr. 22.084,
- the 'CHOROCHRONOS' training and mobility research network, ERBFMRXCT960056,
- the ESPRIT4 project: 'Uncertainty, Knowledge Maintenance, and Revision in Geographic Information System', Nr. 27781, and
- the project 'Spatial and Temporal Object Databases', which is supported by the UK Engineering and Physical Sciences Research Council.

However, the COST action will focus on the modelling of inter-organisational, business transactions, rather than on the investigation of the management of spatial and temporal information for object databases, but the action will benefit from the outcome of the mentioned projects.

Law and economics

The perspective of law and rights in real property has recently been addressed by a group of professors in 'Nordic Academic Views on Real Estate and Cadastre' (Mattsson, Sevatdal, Stubkjær, Viitanen, 1999). An extensive presentation in English of the Swedish property markets appeared in the series European Urban Land and Property Markets (Kalbro and Mattsson, 1995). Similarly, the Finnish market is described in (Viitanen, Vuorio, Yli-Laurila, and Anttila, 1997). In the 1970s the computer-inspired formalisation was reflected in legal circles. The distinction by Eckhoff and Sundby between different types of rules and norms (rules of behaviour, rules of authority, stated norms, internalised norms), and their treatment of 'legal systems' still needs to be applied more widely (1975). Strömholm treats the 'legal system' in a historical context, presenting the efforts of systematising the legal rules, and thus provides a basis for understanding the differences between continental European civil code and Anglo-American common law (1974). These different legal structures are both present in Canada, and are researched at the Land Law Lab of the Centre for Research in Geomatics, Université Laval (Québec), and the Centre for Property Studies, University of New Brunswick.

The concept of property rights has different notions with respect to legal theory and economic theory (Sevatdal, 1999). Departing from works of Douglas C North (1990) and W R Scott (1995) he suggests a set of concepts, including 'institution', which support the analysis and understanding of ownership, tenure, and public regulation of real property in a state. Similarly, the role of the individual and the rational choice is at stage, e.g. in 'Aristotle, Menger, Mises: An Essay in the Metaphysics of Economics' (Smith, 1990). Bo Gustafsson's review of North's theory of institutional economic history (1998), and North's reply (1998) provides a rich set of research questions, some of which may be at least partly addressed by the present COST action, e.g. a categorisation of rules and a specification of institutional functions. The categorisation of rules, suggested by North and Gustafsson differs from the one suggested by Eckhoff and Sundby. The case of real property transactions might provide a testbed for an analysis of the rules issue. Regarding the specification of institutional functions, it seems that cross-jurisdictional comparative analyses of the paperwork that accompany property transactions further the specification of such functions. A comparative analysis of procedures performed to subdivide a unit of real property, suggested tentatively the following functional objectives (Stubkjær, 2002):

- reorganise the rights in the plot and its surroundings at the wish of the parties,
- without compromising the rights of passive (and active) holders of rights,
- in compliance with spatial, environmental and agricultural legislation, etc, and
- maintaining the clarity and efficiency of registration, including establishing of systematically identified plots of land.

At the Fifth Annual Conference of the International Society for New Institutional Economics in 2001, Benito Arruñada presented a paper entitled: The Enforcement of Property Rights: Comparative Analysis of Institutions Reducing Transactions Costs in Real Estate. Rephrased in the non-economic terminology of the present paper, Arruñada explains in economic terms how the institutions of real property rights operate. Land Registries make visible the encumbrances, which the prospective buyer cannot observe on the estate, and the rule-based paperwork effects that 'private contracts (governed by parties' free will) have the consent of the holders of any affected real right (in a process governed by independent officials).' The performance of systems 'is shown to depend substantially on the coherent design of each system' (2001, Abstract).

Among property rights, real property rights are not the only ones. The International Society for New Institutional Economics framed a call for collaboration on 'The Economics and the Governance of Intellectual Property Rights' (ISNIE, 1999). Copyright to geographic information is surely an issue (cf. Meixner and Frank, 1997), but not directly addressed by the present project.

The research issue: Modelling Real Property Transactions

The above review of state of the art suggests that New Institutional Economics make a dominant part of the scientific context of the COST action. In Bo Gustafsson's review of North's theory of institutional economic history, the issue of *institution* is raised: 'What do institutions do? They perform various functions' (1998: 11). The COST action aims at *preparing an exhaustive specification of institutional functions within the domain of real property rights, as it is practised in Europe.*

This includes the specification of the *organisations* who are involved in real property transactions, among others financial institutions and governmental bodies in charge of property taxation, fees, etc., in addition to the organisations mentioned in the introductory sections. Furthermore, it includes an elaboration of the *rules* that govern selected property transactions, encompassing legal prescripts at all levels of detail, professional codes of conduct, and conventions. Due to the resources available, the procedures related to enforcement mechanisms, e.g. compulsory sale, probably have to be postponed beyond project duration. Generally, the specification of functions is not equivalent with a specification of all kinds of transactions or administrative procedures related to real estate. Rather, the assumption is that it can be empirically tested that any kind of transaction or procedure accomplishes one or more of the specified institutional functions.

A view of the domain of real property rights may serve as structure for investigation of the elements of the domain.

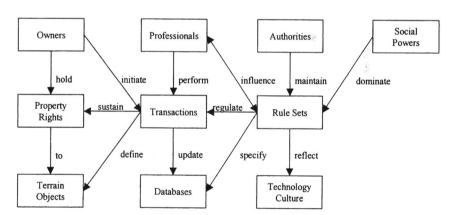

Figure 1: A view of the cadastral problem domain

The specification of institutional functions is established by cross-national (in fact cross-jurisdictional) comparative analyses. However, comparative analyses are hampered by difficulties in translation of language. To overcome these, the project proceeds through semi-formal specification of selected transactions, formalisation of these by means of a language like *Unified Modelling Language*, and finally establishing core *ontologies*. Modelling is a central activity in information systems development (reviewed by Sumrada, this volume). The modelling activity of the COST action includes the description of legal procedures as perceived by the involved professionals and civil servants (independent officials, in the terminology of Arruñada).

An *ontology* is an explication of some shared vocabulary or conceptualisation of a specific subject matter. Domain experts, e.g. those who teach the subject matter at university level, have to co-operate with knowledge engineers, in order to use this formalism and the corresponding computer tools. This is supported still more by the research on *legal ontologies* (reviewed in Visser and Schlieder, this volume), suggesting the following main concepts of legal reasoning: norms, acts, and legal modality.

The notion of acts distinguishes between physical acts and institutional acts: An institutional act, for example a transaction, is a legal qualification of a physical act. For example, the physical act of writing (one's name) on a piece of paper (a deed) establishes the institutional act of granting right of way (assuming that no further demands are prescribed, e.g. the presence of witnesses or a notary). The legal qualification is established through norm and legal modality. Legal ontologies must include an elaborated specification of norms of different kind. A crucial research issue is that *the ontology formalism provides for the specification of legal procedures and the contribution by these procedures to the above-mentioned functional objectives.*

Besides the legal ontology, other ontologies are needed to describe the domain in a distinct way: Deeds, together with maps and markers in the field constitute a set of information carrying, physical media. Human acts: Measuring, observing, communicating, paying in cash, and other activities constitute the physical basis for legal qualifications, captured by the legal ontology. Agents, who performs the acts, specific situations, which provide the contexts for the acts, and the spatial domain and its relation to the representation of it, each needs to be represented in an ontology.

The comparative analysis of the existing, cross-organisational transactions and the databases regarding real property provides, together with the developed core ontologies, *a conceptual framework for pan-European dissemination of public information through portals, and for future information systems.*

Finally, the investigation on the amount and costs of property transactions will take place, when the modelling effort has been used to delineate the transactions from related social activities. The comparative analysis is followed by an explorative analysis of the causes of economic efficiency. Has the efficiency of property transactions changed recently, and for what causes? Has public participation in spatial planning a bearing on property transactions, e.g. by enhancing transparency of the land development process? These and similar questions may be rephrased into proposals for further research.

Methodological considerations

The main effort of the action is to describe, in an objective way, the transactions regarding property rights, that is: routine, ongoing activities with legal implications. Using the land surveyors' measurement of a residential house as an example: The task is not to set out a house to be built or to measure a house to be extended. The task is to measure the house as it is. However, this is only a trivial task on the surface. According to conventions, the surveyor measures the projection of the outer surface of the walls on a levelled plane, (which is not visible), and not along the oblique lines of the terrain, which are visible. The objective measurements of the surveyor imply a deviation from what seems most obvious.

The modelling effort of the action attempts to be as objective as the land surveyors' mapping of a house. However, the modelling effort includes a normative or creative element, when you select the model that 'fits best' with evidence provided by different countries. An answer to this problem seems to be to search for basic regularities, in analogy with the sentence that the shortest distance between two points is measured along the line between them. A more feasible application of this principle is achieved by shifting from the geometrical domain to the linguistic: Among two models, the better is the one that uses the most frequently used words of the language, provided that these words are sufficiently distinct. Also, models should include relevant *core words* of national languages, that is the minimal set of words that every language user has to know. Furthermore, among two models, the better is the one that relate core words with the most abstract, domain specific terminology in the most elegant (simple, consistent, well-structured) way.

Depending on previous efforts in the field of modelling within the participating countries, the following activities is foreseen for few, selected countries:

- Description and comparison of the national variety of forms of land tenure in a way that relates to the major transaction types, and description of the information content of transaction types (conveyance of title, mortgaging, as well as subdivision, reallotment, etc.), as well as of updating information flows.
- Quasi-formal modelling based on the above investigations
- Establishment of taxonomies of technical terms. Provision for semantic translation between different datasets made to support the following comparative analysis.
- Development of formal methods that are feasible for modelling property transactions with a national scope, and ontology eliticing.
- Assessment of the economic efficiency of the transaction processes.

The descriptions are based on studies of literature and occasional visits and interviews, among others to clarify the operation of the ever-developing technical systems, to assist in the application of ontology tools, and to discuss economic assessments. Descriptions are circulated between participating countries, with a view to increase completeness and correctness from a legal point of view.

The formalisation effort includes a survey of available techniques and facilities, and an assessment of different approaches from the point of view of an academic teacher. Techniques, which appear useful in a teaching process, should be given high priority. Several participants of the action research are expected to learn the formalisation techniques as part of the action, and to develop teaching material during this process.

The assessment of the efficiency of different real property systems will be based on the development of an economic model of the systems, which contain the most important variables that determine the resource costs of the various systems. These variables will be derived from the analytical apparatus of transaction costs and property rights economics, as well as from case studies. Efficiency differences will be identified by comparing the importance of some of the variables along with differences in their values. Data on the values of the important variables will be obtained from publicly available databases, from interviews, and from observations. The validity of the operationalisation of the theoretical constructs in the transaction costs and property right theories will be assessed by means of discussion with relevant colleagues and interviews with informed respondents.

Conclusions

The COST action G9 *Modelling Real Property Transactions* has been described with reference to its societal setting, as exemplified by the organisational structure of the Commission of the European Union. State of the art of relevant fields has been reviewed: Land management, formalisation of information, and law and economics. The presented objective of the action is developed from the terms of reference of the action, the Memorandum of Understanding with its Technical Annex. However, the research issues are spelled out, drawing upon the experiences gained during the start-up of the action.

The COST action aims at preparing an exhaustive specification of institutional functions within the domain of real property rights, as it is practised in Europe. A set of core ontologies is to be established by means of computer-supported methodologies. An ontology formalism is envisaged that provides for the specification of legal procedures, and the contribution by these procedures to the above-mentioned functional objectives. A conceptual framework for pan-European dissemination of public information through portals, and for future information systems can thus be established. The comparative analysis is followed by an explorative analysis of the causes of economic efficiency.

The above mentioned modelling effort implies a certain normative or creative element. Linguistic criteria are suggested as means for objectivising the models as far as possible.

References

Arruñada, B. (2001),'The enforcement of property rights: Comparative analysis of institutions reducing transactions costs in real estate', *Fifth Annual Conference of the International Society for New Institutional Economics.*

Bittner, S, von Wolff, A., Frank, A. U. (2000),'The structure of reality in a cadastre' Brogaard, B. (ed) *Rationality and irrationality.* Contributions of the Austrian Ludwig Wittgenstein Society, 8. Austrian L. Wittgenstein Society, Kirchberg, 88-96.

Bundesnotarkammer (2000), 'Europäischer Kodex des notariellen Standesrecht's (Übersetzung aus der französischen Sprache).

COM (1998) 585. Public sector information: a key resource for Europe - Green Paper on public sector information in the information society.

COM (2001) 607. eEUROPE 2002: Creating a EU framework for the expliotation of public sector information. p. 16

COM (2002) 72. 'The Commission's Action Plan for skills and mobility'. p. 37.

COST MoU (2001) 328/00 'Draft Memorandum of Understanding for the implementation of a European Concerted Research Action designated as COST Action G9 'Modelling Real Property Transactions'. COST secretariat 29. January 2001.

Dale, P., McLaughlin, J. (1988), 'Land information management – An introduction with special reference to cadastral problems' in *Third World countries.* Clarendon Press, Oxford. p. 266.

Dale, P., McLaughlin, J. (1999), 'Land administration', Oxford University Press, Oxford.

Dale, P., Baldwin, R. (2000), 'Lessons learnt from the emerging land markets in Central and Eastern Europe'.

de Soto, H (2000), *The mystery of capital – Why capitalism triumphes in the West and fails everywhere else* Bantam Press, London. p.243.

EC (1998), Nine infringement cases relating to freedom to provide services against seven Member States (3 July 1998).

EC (1999), Composite Paper on the Commission Reports 1999, October 13, 1999.

ECE (1996), Economic Commission for Europe: Land administration guidelines with special reference to countries in transition. United Nations, Geneva. p.94.

Eckhoff, T., Sundby, N. K. (1975), *Rettssystemer: Systemteoretisk innføring i rettsfilosofien Tano,* Oslo. p. 272.

ETeMII (2001). Reference Data White Paper (v 1.1) IST-1999.12096 p. 37.

European Mortgage Federation (2001) European Agreement on a Voluntary Code of Conduct for Pre-contractual Information on Home Loans.

Frank, A. (1996),'An object-oriented, formal approach to the design of cadastral systems', Proceedings of *7th International Symposium on Spatial Data Handling,* SDH '96, Advances in GIS Research II, (M. J. Kraak, amd Molenaar, M., eds.) Delft, The Netherlands (August 12-16, 1996), Published by IGU, Vol. 1, pp: 5A.19 - 5A.35.

Goetschy, J. (1999), 'The European Employment Strategy: Genesis and development', *European Journal of Industrial Relations.* 5 (2):117-137.

Gustafsson, B., 1998. Scandinavian Journal.

Holstein, L. (1996), 'Towards best practice from World Bank experience in land titling and registration', Paper, presented at *International Conference on Land Tenure and Administration,* Orlando, Florida, November 1996. p. 26.

ISNIE (1999), 'The economics and the governance of intellectual property rights' Newsletters January/February.

20   *The Ontology and Modelling of Real Estate Transactions*

Kalbro, T., Mattsson, H. (1995), 'Urban Land and Property Markets in Sweden', *European Urban Land & Property Markets* 5. UCL Press, London. p.193.

Kjellson, B. (2002), 'What do Americans pay for not having a public LIS?', *FIG XXII International Congress*, Washington, D.C. USA, April 19-26 2002.

Larsson, G. (1991), 'Land registration and cadastral systems', *Tools for land information and management*. Longmans Scientific, Harlow.

Larsson, G. (1997), 'Land management - Public policy, control and participation', *Byggforskningsrådet*, Stockholm. p.232.

Lemmen, Oosterom, P. Special Issue: Cadastral Computers, *Environment and Urban Systems* 50.

Longley, P. A., Maguire, D. J., Goodchild, M. F. and Rhind, D. W. (Eds) (1999), *Geographical Information Systems*. 2nd edition, Vol 1-2. John Wiley. New York.

Mastricht Protocol, 1992. Protocol on the acquisition of property in Denmark.

Mattsson, H. Sevatdal, H., Stubkjær, E. Viitanen, K. (1999),'Nordic Academic Views on Real Estate and Cadastre' *Kart og Plan* 59 (3) International Edition: Real Estate and Cadastre.

Meixner, H., Frank, A. U. (1997), *Study on Policy Issues Relating to Geographic Information in Europe*. EC 1997.

North, D. C. (1990), *Institutions, institutional change and economic performance*. Cambridge University Press, New York.

North, D. C. (1998) Reply.

Office of the UK Deputy Prime Minister (1998), 'Housing Market Transactions: International Comparisons', *Housing Research Summaries* No. 77.

OJ, (1999) *Notice of open competition*. C 122, 04/05/1999 P. 0035 – 0035.

OJ, (2001) *Fortgang der Erstellung eines Katasters für Griechenland*. C 318 E 13/11/2001 S. 0076 – 0077.

Scott, W. R. (1995), *Institutions and Organizations*. SAGE Publications.

Sevatdal, H. (1999), 'Real estate planning: An applied academic subject', *Kart og Plan* 59 (3) 258 - 266.

Silva, M. A., Stubkjær, E. (2002), 'A Review of Methodologies used in Research on Cadastral Development Computers', *Environment and Urban Systems* 51 (5) .

Smith, B. (1990), 'Aristotele, Menger, Mises - An Essay in the Metaphysics of Economics', *History of Political Economy Annual supplement* 22, 263 - 288.

Smith, B. (1995), 'On Drawing Lines on a Map', pp 475-484 in A. U. Frank, W. Kuhn and D. M. Mark (eds.), *Spatial Information Theory*. Proceedings of COSIT '95, Berlin: Springer.

Smith, B., Zaibert, L. (1997), *The Metaphysics of Real Estate* Buffalo, NY.

Steudler, D., Williamson, I. P., Kaufmann, J. and Grant, D. (1997),'Benchmarking Cadastral Systems', *The Australian Surveyor* 42 (3) 87-106.

Strömholm, S. (1974), 'Det juridiske systembegreppets uppkomst och utvickling',*Tidsskrift for rettsvitenskap* (3) 225 - 244.

Stubkjær, E. (1999), 'Cadastral research - Issues and approaches', *Kart og Plan* 59 (3) 267 - 278.

Stubkjær, E. (2002), 'Modelling Real Property Transactions', *FIG XXII International Congress*, Washington, D.C. USA, April 19-26 2002.

Viitanen, K., Vuorio K, Yli-Lauria, M., Anttila, K. (1997), 'Urban Property Market and Land Law in Finland', B 80. *Espoo*. p. 76.

Visser, U, Stuckenschmidt, H., Schuster, G., Vöegele, T. (2002), 'Ontologies for geographic information processing', *Computers & Geosciences* 28 103-117.

Williamson, I. and Fourie, C. (1998), 'Using the Case Study Methodology for Cadastral', *Reform. Geomatica* 52 (3) 283-295.

Zevenbergen, J. (1998), 'The interrelated influence of the technical, legal and organisational aspects on the functioning of land registrations (cadastres)', *FIG XXI International Congress*, Commission 7, p. 130-145. Brighton.

Chapter 2

Aspects of Real Property Rights and their Alteration

Hans Mattsson

Abstract

It is essential to divide land into spatially defined rights, such as property units. Those rights should be basically adapted to economic and other activities in the society for which they are created. The result can be quite complicated relation between rights such as ownership, leases, easements etc. Furthermore, it is important that the rights should be amenable to change. The processes for bringing about these changes should be designed so as to be easy to handle and so as not to involve unnecessary expense, otherwise there is a risk of changes being obstructed and the economic potential of the land not being fully utilised in relation to what is possible. This article deals with aspects of real property rights and processes for changing such rights.

Division of land into property rights

Land is fundamental to human activities, but in order for its use to be well organised, provisions are needed, otherwise chaos threatens. The provisions can decide who may use the land, what may be done with it, who may not enter it, and so on. In modern societies, questions of these kinds are regulated by means of laws and other statutory instruments, even though custom may also be of relevance.

Looking at the states of Western Europe, one finds that land is divided into rights which can belong to persons, business undertakings, organisations, local authorities, the State, and others. The instruments governing rights are extensive and, moreover, differ partly from regulations governing other property. This distinction is reflected by terms like immovable/movable property and personal/real property. The right to land may take the form of ownership, but there also exist rights of other kinds which are governed by law – the right of user, for example.

Fundamentally, a right entitles one or more persons to use the land while others are excluded from doing so. The land is individualised. The reason for this individualisation is a number of problems which society has been forced to solve. One factor frequently referred to is that open access entails over-exploitation if the asset has a value, i.e. if everyone has access to the resource. Hardin (1968) sums this all up as 'The Tragedy of the Commons'. Ostrom (1990) refutes a great deal of Hardin's arguments by reviewing different forms of common property and showing the conditions which, in principle, have to exist in order for common property, though without open access, to work or not to work. Thus the line of demarcation for individualisation is not between common and private property but between open access and use founded on exclusive rights.

Individualisation is a way of guaranteeing the user or users the production from a piece of land and excluding others from the use of it. This is prompted by some form of user motive, but perhaps also by a fiscal motive. The State requires a distinct object – the land – on which to levy tax. What is perhaps even more important in western societies, however, is individualisation as a precondition of investment security. The proprietor of a right can calculate in advance whether or not an investment will pay.

There are more reasons for individualisation. As de Soto (2000) sees it, clear and simple rules for the transfer and mortgage of clearly defined property rights have in principle been propelling the economic development of our western European societies for a long time. Given smooth-running systems for the transfer of real property, someone perceiving a profit potential which someone else does not see or does not wish to develop can offer to take over a property or a right in it with a view to exploiting that possibility. Cheap systems for transfers of rights make more profit opportunities available for exploitation than when expensive, complicated processes have to be used. If the rights can be mortgaged without too much trouble, they can also contribute toward the development of the capital market. Capital tied up in real property can be released for investments by means of loans on the security of the right.

The structure of rights in real property depends on terms of production and on the innovative capacity of the legislature, but also on historical tradition, legal tradition included. This article considers what aspects constitute real property, how different rights in properties can be organised in relation to each other and the need to transfer and also to change the rights in a society with a developing economy.

Real property in the light of Swedish examples

It can be of interest to consider what real property is. This is not altogether easy to pin down, as will be seen from the following description, based on Swedish conditions. Similar difficulties probably occur if some other country is taken by way of example.

It has to be mentioned that real property in Sweden is regarded as immovable property in contrast to movable. The distinction between rights in rem and in personam is of little relevance and ownership as well as leasehold are enforceable against all third parties (Johansson, 1998). Real property suggests something real – solid, permanent – and, ultimately, land.

Land in Sweden is divided into property units with unique registration numbers and such a unit can consist of one or several parcels. Buildings and other things connected to the land by its owner are fixtures belonging to the property unit. In order for a fixture to be detached from the real property, the new owner must remove it, or else the fixture must be transferred to another property but remain in the same location, subject to a right resembling an easement, and then only by order of a public authority. Accordingly, a house built on somebody else's land by virtue of a leasehold contract is considered personal – movable – property in Sweden.

All land in Sweden is divided into property units with different owners. The owner may be the State, a municipality, a private person, a company, a foundation, a church and so on. Furthermore, all property units are registered. Certain areas, though, are joint property units in which other property units have shares. These properties with shares in joint property units may be tangible property units defined on the ground, but they can also be purely notional property units existing only in the sense of having a share in a joint property unit.

Water areas in lakes (with the exception of certain of the larger lakes) and up to a certain distance out to sea (there are several rules governing these delimitations) also constitute property units or joint property units. A water area of this kind can be included in an on-shore property unit, but it can also be quite separate, consisting of the water area alone.

The property owner cannot claim ownership of air (atoms drifting by in the wind), neither to water in the ground nor water in an area which he owns in a lake or in the sea. He merely has the right of using the air and water. Where the water is concerned, moreover, he must have permission for extraction in excess of domestic needs. Nor do extractable 'concessionary minerals' belong to him, even though they are solid and permanent, to say the least of it. In principle, all economically interesting minerals apart from gravel and soil are defined as concessionary minerals.

Fishing rights can be completely separated from the title to the water area. Thus the right to fish may be owned by one person and the water area by another, while the water itself belongs to nobody. In cases of this kind, the fishing right is regarded and registered as a property unit and treated according to the rules on real property, even though the fishes are not fixed and stationary.

Conditions in Sweden can be summed up by saying that real property is usually solid, fixed and permanent – 'concrete' in one sense or another – but that, in the ultimate analysis, real property is what the law defines it as being. It is the law that decides, not the physical characteristic, even though the latter is usually decisive. The reason for this way of looking at things is partly to be found in legislation and case law, but also in the economic facts of agrarian society in days gone by. These matters are so complicated that they defied all efforts of the legislature to arrive at a comprehensive definition of real property (Prop., 1969; Prop., 1970).

In this connection there are a number of aspects of ownership which have to be considered. The nature of ownership has engaged the attention of a succession of lawyers and philosophers. Snare is one of the writers highlighting the right to use an object, exclude others from doing so, and transfer it to another owner (Snare, 1972). Ownership, moreover, is distinguished from the right of user by the chronological aspect, ownership being, in principle, perpetual, whereas a right of user is of fixed duration, reverting eventually to the grantor.

Another important aspect of the ownership of real property is its frequently negative definition in law, e.g. in Sweden. The main principle is that legislation does not address the powers vested in the owner of land; instead it deals with the powers which he does not have. On the other hand the law entitles the owner to grant clearly defined rights in his real property to another, e.g. in leasehold form. In other words, the grant takes the form of a positive definition.

Some interesting ideas on ownership of real property have been put forward by Bergström (1956). He begins by observing that a property unit has an owner but that this right is restricted by two circumstances. Firstly, as a result of statutory rules, there are certain rights which the landowner does not have at his disposal, because they have been socialised in some respect. Undén (1928) argued long ago that characteristics of this kind could not be viewed as a part of ownership, but Bergström maintains that they are a latent right of ownership, whereas other characteristics of a property unit are included in the current right of ownership. If the statutory provision is repealed, the right of using the characteristic accrues to the property owner, i.e. to the proprietor of the area defined on the ground. The other kind of restriction on the right of ownership occurs when the landowner grants a certain right in his property, voluntarily or otherwise, to another. The right of ownership concerned is limited in positive terms. What, then, is the ultimate core of ownership? If the entire property unit with all its rights has been granted to somebody else, the owner is really only entitled to the value (the leasehold revenue) plus the possibility of recovering the right when the grant of it expires. The right to the value alone becomes even clearer if the property unit is expropriated. So in the ultimate analysis, according to Bergström, perhaps the core of ownership is merely the right to a value generated by a property unit. One can add that this is at least the approach in the mortgage market.

We may add that in Sweden, as in many other countries, rights of user are regarded as personal property, though they have many characteristics in common with real property. The type of property to which different rights are referable is ultimately a matter for legislation. Perhaps the same can be said concerning the ownership of real property. Ownership is established, directly or indirectly, through legislation and is not absolute or self-evident. Conditions in the Soviet Union, at least, demonstrated that private ownership of land can be non-existent.

Land characteristics and rights

One and the same area in a town can, for example, be made to serve as a park, a building site or a street. The ground beneath can be used for garaging, an underground railway, or water and sewerage mains. A rural area can, for example, be applied to agricultural production, timber production, hunting or occupied by power transmission lines. Underground mining activities can occur beneath both urban and rural lands. Certain activities can be concurrent, while others are mutually exclusive. Thus, depending on productive capacity and other natural qualities, as well as location, land can have a variety of characteristics which are useful to man. Use can be especially complicated in urban areas, as illustrated in Figure 2.

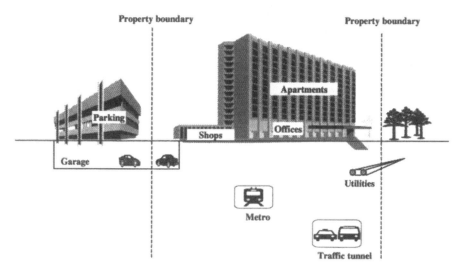

Figure 2: Versatile use of property units (based on Julstad, 1994)

If, then, we consider a piece of land from the viewpoint of production and consumption, it has a host of potential uses. Let us call them the attributes of the land. One and the same person may have the title in perpetuity to all attributes which an area gives rise to. This is what we might call absolute ownership. No one is entitled, without the landowner's consent, to do anything with or to the property, whereas he himself can do what he likes with it without asking anyone. The only possibility for anyone else to be allowed to use the land is by obtaining permission from the owner, or alternatively being allowed to take over the ownership of the land. Absolute land ownership, however, is hard to imagine in a society, because activities on the land often have consequences for other people. Usually, therefore, all landowners incur some form of regulation, with the result that absolute ownership comes to comprise a latent and a current component. The boundary between these two components can always be changed, but full transition to an absolute right seems an unlikely possibility.

Taking the attributes of the land as our starting point, and assuming a system of ownership in which rights are also to be transferable to others in the longer or shorter term, a system of rights has to be constructed which, in order to be viable, will of necessity be quite complicated. A system of this kind is illustrated in Figure 3.

Initially the land has a number of attributes which are then legally processed in various ways. Certain attributes can be available to all comers, i.e. access to them will be open. Examples of such open access can be quoted from Sweden. If an area is undeveloped, e.g. has not been built on and is not being farmed either, the general public have free access to it by virtue of 'Everyman's Right' (allemansrätt), as the term goes in Sweden, meaning a universal right of access. This entitles people, for example, to walk over the land, pick berries and mushrooms on it and camp on it for a single night. If it is a water area, the general public are entitled to cross it by boat. In short, open access applies.

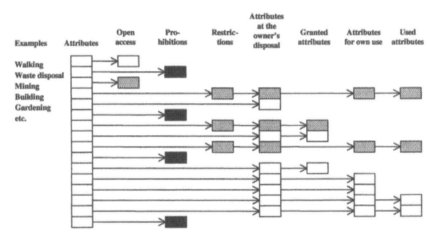

Figure 3: Attributes and rights in real property (black=not possible to use; grey=restricted use; white=possible to use)

Another form of open access which changes into an individual right can be instanced with the right in Sweden of extracting what are termed concessionary minerals. Anyone is entitled, subject to permission, to investigate mineral deposits in an area, and if the area proves economically interesting from the viewpoint of mineral extraction, they can also obtain a fixed-term exploitation permit. As the law now stands, the above-ground property owner is not entitled to any compensation, so long as the use of his property is not otherwise affected. Here we have open access giving way to a right segregated from the ownership of the land. This approach is prompted by a desire on the part of government for metalliferous deposits to be exploited in the interests of the national economy. The permit procedure, however, ensures that environmental implications can be taken into consideration and that whoever obtains permission to extract minerals is economically capable of accomplishing the project.

The State or other decision-making bodies can, through the exercise of statutory powers, prohibit activities. Other activities can be regulated, meaning that a certain activity may not be undertaken until permission has been obtained. Suppose there is no automatic right of building. Not being allowed to build amounts to a prohibition. Permission, if granted, may be subject to restrictions, e.g. concerning design.

Regulatory instruments notwithstanding, there remains a number of attributes of the property which the owner can make use of, even if he needs permission in certain cases. These attributes can be said to be at the landowner's disposal. Both freely usable attributes and those which are subject to permission can, however, be granted to another, i.e. the landowner can surrender the use of them. For example, he may grant a lease for the building of a weekend cottage (restricted use) or the hunting rights on the land. A number of attributes then remain which can be used by the landowner, but these may come into conflict with other rights which are used, or else be of no interest to the landowner. Of all the land's attributes, perhaps in practice only a few remain which are used by the landowner.

Figure 3 starts with attributes of an area of land. Legislation then makes it possible to create rights relating to those attributes, but the rights must be definable in such a way as to be viable in terms of both scope and content, otherwise we will find ourselves with a legal system which inhibits the development of society, economic development included.

The definition of desirable and undesirable is ultimately a political issue, since legislation is enacted by parliaments or similar assemblies. In the competition to which rights are often subject, systems may be developed which are more or less functional, and so we also have to bear in mind the question: functional and not so functional for whom? A non-functional system can be illustrated as follows. In a system of private ownership of urban land which, theoretically, extends to the centre of the earth and affords no possibility of granting space below ground, a municipal underground railway cannot possibly be constructed on/under land which is not municipally owned. In principle, therefore, the development of an underground railway system may be impossible – that is to say, unless legal procedures are created for gaining access to areas below ground.

Figure 3 can readily be viewed as expressing bundles of rights, i.e. a system of rights in which different rights in the same area are clearly distinguished, but with different owners or users and with the rights themselves transferable independently of each other. In systems where the right to property is conceived of as a single whole, this view, in principle, is not tolerated, but this does not prevent the bundles-of-rights approach being made a foundation for studies of legislation. Given the possibility of a number of attributes being used simultaneously and by different holders, a breakdown of the system of rights into components can be an approach to understanding a system of rights (for explanation of bundles-of-rights, see Barron's, 1984).

We can also take the following, topical example of adjustment of the development of rights to new conditions. A property owner in Sweden owns the space enclosed by the property unit's boundary marks. There is no formal boundary upwards or downwards. In reality the territory owned in these directions is limited by what is practically usable (2D delimitation with X and Y co-ordinates). Since these units do not always fit the needs from an administrative point of view, a number of possibilities have been evolved for granting rights in property units, e.g. land leasehold, rental tenure of homes or offices in a building, public road rights, easements, utility construction rights and the right of constructing a private road to be shared between several property units. Some of these rights are transferable, others are not.

Property units, then, can only be created as 2D units. But this is not always a good solution, despite the possibilities of creating other rights in properties. Sometimes a purely horizontal delimitation is desirable for purposes of maintenance, mortgaging and risk assumption. For this reason the Swedish Government is currently drafting legislation making it possible for property units to be defined on the horizontal plane (3D property units with X, Y and Z co-ordinates). In this way the spatial extent of property units can be adapted more closely to actual needs, and awkward arrangements involving easements or rights of user will in certain cases be avoidable (Julstad, 1994; Julstad and Ericsson, 2001).

The need for land law to support dynamism

As we have now seen, there are a number of reasons for individualising rights attaching to land. But if those rights are individualised, other problems arise to which legal solutions have to be found. In a society with static rights, i.e. rights which, once established, cannot be altered, rules are only needed for establishing rights once and for all and subsequently defining their scope in the event of disputes. But the very necessity of rights being transferable shows that legislation must also include dynamic components, i.e. components whereby smooth-running changes of rights to land are made possible.

It is a fact of nature that individual ownership cannot subsist in all perpetuity. Rules of inheritance have to be developed. A general obligation of possessing a property for life is no rational solution either, because when a property is no longer needed, the rational thing is to dispose of it. In this way it can pass to someone who has more need of it.

The possibility of transferring the property directly between two parties, e.g. by putting it on the market, enables the seller to transfer the capital present in the property from one place to another. He sells in one place and buys in another. If, however, it is very expensive to transfer the capital tied up in one property unit to another, or to dispose of the property unit altogether, the sale may be inhibited. Suppose, for example, that someone wishes to move from one locality to another because he has been offered a better job and therefore has to sell his home in order to procure a new one, of a similar kind, in the new locality. Assume, for the sake of simplicity, that property prices are the same in both localities. If the cost of selling and buying is high because of estate agents' commissions, notarial charges, stamp duties, opening charges for credits, land registration charges and so on, there will be less incentive for moving than if these charges are low. The balance between advantages and disadvantages decides. It is also important for trade, offices, industry and other activities to be able to relocate smoothly, without excessive transactional costs, and also for farm and forest properties to be able to change hands.

Thus the design of systems for the transfer of real property makes an important difference to a society's economic efficiency. That importance is not confined to the property market as such. It also has implications for the economy in general, as regards the willingness of people and undertakings to relocate. The system ought hardly to be constructed so as to unnecessarily impede transfers, otherwise existing property owners will generally hang onto their rights too long, with the result that others, capable of using those rights more efficiently, are prevented from taking over.

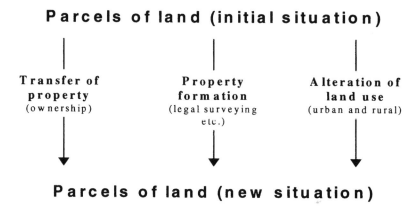

Parcels of land (initial situation)

| Transfer of property (ownership) | Property formation (legal surveying etc.) | Alteration of land use (urban and rural) |

Parcels of land (new situation)

Figure 4: **Necessary procedures in land law for change (based on Mattsson, 1997)**

Smooth-running systems for transfers of existing property units, however, are not all it takes to support dynamic development. Spatially inappropriate property units can also be a drag on development. If, for example, a factory in need of expansion occupies 100 per cent of the land area of a property unit at the same time as neighbouring land belongs to someone else, there have to be methods for extending the existing property unit onto the neighbouring land. Price and the neighbour's general willingness will, however, decide whether the factory can expand in its present location or instead will have to carry on as it is, expand in an upward direction, improve the logistics of its production or move elsewhere.

Readjustment of a property unit can be viewed as the transfer of an area of land in this particular instance, but in principle this is the wrong way of looking at things. The physical design of property units is one thing. Transfers of them are another. Transfers can be used as a readjustment tool, but property readjustment need not necessarily involve any transfer between owners. One and the same person, for example, may own both the factory property we have been talking about and a neighbouring property – a farm, for example. He may want to adjust the boundary between his property units, so as to qualify for a bigger mortgage on the factory property while refraining from mortgaging the agricultural property. Ownership of the land, in principle, is unaffected by the land transfer, since he owns both properties, but in practice these are two different legal units which can be separately dealt with. There are a number of property formation methods for changing the extent of property units, e.g. subdivision, amalgamation, reallotment, land readjustment, land consolidation and partitioning. Systems of property division and readjustment are needed which will not be unnecessarily expensive, so that desirable changes will not be frustrated by considerations of expense.

Experience, not least from the infancy of industrialism, has taught us the importance of regulating land use so as to prevent external effects, such as a landowner polluting a neighbour's land or blocking his scenic view of the water. An activity, in other words, can destroy values to other parties. Infrastructure co-ordination, often with the aid of spatial planning, is another reason for exerting control through the award or refusal of permits.

The normal practice in Europe is for national and local government, acting through political assemblies or public authorities and exercising powers conferred by legislation or statutory instruments, to regulate land use. In our example of the factory, it might be possible to expand upwards or downwards instead of sideways. If this is not possible without a permit, then the sanctioning authority regulates the content of the right within a property unit, as regards the use which can be made of it. But if a permit is needed for alteration of use, the regulatory instruments have to be framed so as not to impede change as such but only phenomena whose disadvantages outweigh their benefits. In particular, permit systems with many, mutually independent decision-makers basing their decisions on vague criteria can have a devastating impact on development. For the less wealthy as for the uninformed, the expense and risk entailed even by starting a permit application procedure will be prohibitive.

Three important aspects have been highlighted, indicating the need for systems of change relating to property units. Mattsson (1997) has illustrated the processes of change graphically (Figure 4). That figure, referring to the ownership of real property, points to the need for processes to change the ownership (transfer), the design of property (property formation) and the permissible use (alteration of use).

Even though, in principle, three different phenomena have been highlighted in Figure 4, combinations are of course possible. The acquisition of an area of farmland for the building of a home can include the activities of purchase, subdivision and building permission. In other words, all three components are included. But in other cases they can be separately dealt with, as for example in the case of separate purchase of an existing property unit (transfer), separate subdivision of one's own land without any sale (property formation) and application for permission to build on an existing forest property of one's own (alteration of use).

On the basis of the arguments concerning Figure 3, the property owner's current right of ownership can be limited by granting rights in the property. The three mechanisms of change, therefore, must also be available to make possible changes in grants of rights in a property.

Political control of the three processes of change has not been highlighted except in connection with the alteration of land use. This control can apply to all three processes and can be exercised directly, through political decisions, but also indirectly, through examination of the merits leading to administrative and judicial decisions. Control can both facilitate changes and impede them.

Figure 4 has not highlighted the mortgage system. The surrender of a property as security for a loan is a form of transfer. But the transfer of the pledge and getting the property back when the loan is paid is not rational in capital-intensive economies with high real estate prices and costly facility investments, where the purpose of the loan is to facilitate the purchase of property and/or investments in it. Capital is released by creditors to be employed in the property owner's activity. But it is important to note that the activity need not be attached to the property. Even so, the mortgage is a right (the right of foreclosing on the pledge if the borrower defaults on his liabilities) which has been granted in the property, and mortgages can therefore be seen as a form of granted right, in the same way as grants of other rights.

The degree of security in the property law system, i.e. for all the activities described in Figure 3 and Figure 4, including the possibilities of mortgaging property, depends on the clarity of the legislation, but also on the case with which reliable particulars can be obtained concerning a property and, ultimately too, the possibility of realising securities for loans. Universally available particulars facilitate trade and mortgaging, while concealed or elusive particulars impede them. Real property registers containing particulars of owners, the structure of the property, appurtenant rights, charges etc., have an important part to play here, and the ease or difficulty of handling transfers will be commensurate with the reliability and accessibility of such registers. A state-guaranteed ownership register, for example, provides security for purchasers, while the non-existence of records may necessitate the investigation of purchasing chains a long way back in time, perhaps for every new transfer of an existing property unit. True, a system of insurance, as in the US (Simpson, 1976), can be developed to cover situations of this kind, but even that is likely to be based on the insurance companies, either separately or collectively, keeping registers of properties so that they will not need to investigate the chain of purchases for every new insurance occasion. Clearly, then, a properly designed system of property information facilitates changes.

The credit market, with real estate as security, can in fact be viewed as a locomotive of the national economy, as witnessed by the large proportion of bank loans in western societies referable to real estate mortgages. Historically speaking, therefore, mortgage facilities have been one of the driving forces behind the construction of well-ordered systems of property information. Another motive force has been the desire to secure rights of ownership and facilitate the transfer and grant of rights. Fiscal considerations are a third (Larsson, 1991; Dale and McLaughlin, 1999).

Conclusions

Summing up, it is hard to imagine societies changing economically without changes also occurring in the ownership and use of land. Legal systems supporting this are normally complicated affording many opportunities for creating different types of rights. Rights in land are also admittedly slow to change, because investments in them are often for the long term, but flexible systems are needed for different types of changes, otherwise existing structures of rights will become a form of legal brake on development. It is not altogether unreasonable to assume that the faster a society changes, the more adaptable its property-based system of rights will have to be. Perhaps this assumption can also be applied in reverse, by saying that the more flexible the system of rights in land, the more easily social change will come about. But we must beware of oversimplifying matters, because processes of change have to allow for different interests, so as not to endanger the legitimacy of public society. Nor can freedom of contract be taken too far, because standardisation often makes for a cheaper system.

References

Barron's, (1984), *Dictionary of Real Estate Terms*, Barron's Press, Woodburry etc., US.

Bergström, S. (1956), Om begreppet äganderätt i fastighetsrätten, *Svensk Juristtidning*, 1956:10.

Dale, P. and McLaughlin, J. (1999), *Land Administration*, Oxford University Press, UK.

Hardin, G. (1968), *The Tragedy of the Commons, Science* 162:124, pp. 3-8.

Johansson, R. (1998), Sweden, in Hurndall, A. (ed), (1998), *Property in Europe: Law and Practice*, Butterworths, UK.

Julstad, B. (1994), Tredimensionellt fastighetsutnyttjande genom fastighetsbildning. Är gällande rätt användbar?, *Juristförlaget*, Sweden.

Julstad, B. and Ericsson, A. (2001), Property formation and three-dimensional property units in Sweden, in van Oosterom, P. J. M., Stoter, J. E. and Fendel, E. M. (eds) *3D Cadastres - Registration of properties in strata*, International workshop on 3D Cadastres, FIG, Denmark.

Larsson, G. (1991), *Land Registration and Cadastral Systems*, Longman Scientific & Technical, UK.

Mattsson, H. (1997), The Need for Dynamism in Land Law, in *Land Law in Action*, Swedish Ministry of Foreign Affairs and Kungl. Tekniska Högskolan, Sweden.

Ostrom, E. (1990), *Governing the Commons*, Cambridge University Press, UK.

Prop., (1969), Kungl. Majt:s proposition 1969:128 med förslag till fastighetsbildningslag, Sweden.

Prop., (1970), Kungl. Majt:s proposition 1970:20 med förslag till jordabalk, Sweden.

Simpson, S. R. (1976), *Land Law and Registration*, Cambridge University Press, UK.

Snare, F. (1972), The Concept of Property, *American Philosophical Quarterly*, 9 (2), pp. 200-206.

de Soto, H. (2000), *The Mystery of Capital*, Basic books, New York.

Undén, Ö. (1928), Några synpunkter på begreppsbildning i juridik, in *Festskrift tillägnad Axel Hägerström*, Sweden.

Chapter 3

Real Estate: Foundations of the Ontology of Property

Leo Zaibert and Barry Smith

Abstract

Suppose you own a garden-variety object such as a hat or a shirt. Your property right then follows the age-old saw according to which possession is nine-tenths of the law. That is, your possession of a shirt constitutes a strong presumption in favor of your ownership of the shirt. In the case of land, however, this is not the case. Here possession is not only not a strong presumption in favor of ownership; it is not even clear what possession is. Possessing a thing like a hat or a shirt is a rather straightforward affair: the person wearing the hat or shirt possesses the shirt or the hat. But what is possession in the case of land? This essay seeks to provide an answer to this question in the form of an ontology of landed property.

The Boundaries of Landed Property: How Far Does Your Property Extend?

In his far-reaching study of property rights, Richard Pipes discusses the etymology of 'possession' and cognate terms. He tells us:

> Some primates assert exclusive claims to land by physically occupying or 'sitting' on it. This behavior is not so different from that of humans, as indicated by the etymology of words denoting possession in many languages. Thus, the German verb for 'to own', besitzen, and the noun for 'possession', Besitz, literally reflect the idea of sitting on or, figuratively, settling upon. The Polish verb posiadać, 'to own', as the noun posiadłość, 'property', have an identical origin. The same root underpins the Latin possidere, namely sedere, 'to sit', from which derive the French posséder and the English 'to possess'. The word 'nest' derives from a root (nisad or nizdo) signifying 'to sit'. The monarch occupying the throne has been described as engaging in 'nothing else but the symbolic act of sitting on the realm' (Pipes, 1999, 68).

In this passage Pipes correctly emphasizes the 'symbolic' and 'figurative' nature of this 'sitting on' and 'settling upon' the land. For his purposes it is not important to ask how much land a person (or primate) possesses (or owns) by symbolically sitting on it. It is unlikely that the person would be claiming exclusivity only over the surface of the land he is actually touching. Much more likely is it that a person would claim exclusivity over a region much larger than the area in actual contact with his body. And the symbolic practice of sitting gives absolutely no clue as to what the extension and boundaries of the land over which the person is claiming exclusive rights might be. Thus, the object a person claims to possess or to own is not well defined. Note that this factor of indeterminacy or uncertainty in the borders of one's property has no analogue in the realm of shirts and hats. It is geographic in nature.

It is our purpose in what follows to stress the special character that landed property exhibits amongst the many forms of property rights. Understanding this special character will then shed light on what is needed for a more adequate account. Such an account must encompass not only the dimension of law but also those of politics and economics (Stubkjær, 2001). Here we seek to lay bare the foundations of the needed full ontology of landed property.

The Politics of Landed Property: What Can We Own?

The crucial importance for political affairs of landed property (or real estate, we shall use these two expressions interchangeably) has been eloquently summarized by Rousseau:

> The first person who, having fenced a plot of ground, took it into his head to say this is mine and found people simple enough to believe him, was the true founder of civil society (Rousseau, 1992, 44).

There are two aspects to Rousseau's view that deserve special attention; one concerns geography, the other ontology; more precisely the ontology of social reality. First, the act of fencing off need not, in the context of this passage, be restricted to the case where some physical boundary is constructed. It can be seen as including also the establishment of fiat boundaries – for example when you tell people where the borders of your property lie, or when you simply mark its corners (Smith, 2001). To fence a plot of land is to create something new. The land itself, of course, exists before the parcel is plotted, but the act of fencing off nonetheless creates a new object. Second, this act alone is not sufficient for such object-creation. The latter requires also the existence of what John Searle calls collective intentionality (Searle 1995); that is, it requires that other persons (simplemindedly or not) believe that the land is indeed the property of he who fenced it off. Only then can a property right be said to arise.

This means that a comprehensive study of landed property will have three interconnected dimensions:

- a geographic dimension, having to do with the peculiarities of the ways in which real estate is related to the land itself (and thus also with the issue of boundaries);
- a cognitive dimension, having to do with the interrelations between such geospatial phenomena and our culturally entrenched beliefs and convention;

- an ontological dimension, having to do with what real estate is.

We can throw some light on the latter by considering first of all the more general question of what can be the object of a property right of any sort. Let us use the term 'thing' to refer to anything that can in principle be owned. The German legal philosopher Adolf Reinach provides a useful first analysis of this notion, pointing out that:

> The concept of a thing [Sache] in no way coincides with that of a bodily object, even if positive enactments would restrict it to this. Everything which one can 'deal' with, everything 'usable' in the broadest sense of the word, is a thing: apples, houses, oxygen, but also a unit of electricity or warmth, but never ideas, feelings or other experiences, numbers, concepts, etc. (Reinach 1983, 53).

Reinach's passage carries the suggestion that, even though the concept of thing is not to be identified with that of a bodily object, still: things must be concrete. Abstract entities such as numbers and concepts fall outside the range of what can be owned. As Reinach himself would have accepted, however, it is perfectly possible that entities such as computer programs, architectural designs, and so forth be owned. And even leaving aside such issues of intellectual property, we shall see that there is an important further class of abstract entities – rights themselves – which fall within the domain of what is ownable.

Reinach suggests that being 'usable' might be a necessary condition for something's being ownable; but it is not a sufficient condition. There is a long list of objects regarding which it is difficult to say whether they can be owned, though it is clear that these objects can be used in varied ways. Do we own ourselves? We have certain rights over our bodies, but are they property rights? (Munzer 1994, 1995) Whether or not human corpses, body parts, children, can be owned are difficult questions to answer (Ryan 1994). But the difficulties associated with the idea of ownership in such entities are of a different sort from those which arise in the case of land. The limitations which many societies place on the ownership of human corpses stem from religious and ethical views, not, for example, from any difficulty in ascertaining the boundaries of corpses. Similarly, limitations on the right to commercialize our body parts seem to stem from ethical considerations rather than from any ontological difficulty in determining the boundary of, say, a lung. Such a geographic dimension may, though, arise in relation to the buying and selling of fetuses, where we do indeed face a difficulty in determining the boundary between fetus and mother (Smith and Brogaard 2002).

We shall here, however, leave aside the discussion of those objects which are excluded from being ownable as a result of moral and religious views, and concentrate exclusively on the case of ownership in land.

The first step in trying to analyze land as an object that can be owned is to appeal to the age-old distinction between movable and immovable things. Land is the quintessential immovable thing. (The German term for real estate law is 'Immobilienrecht'.) The term 'real estate' refers precisely to those immovable things which are the objects of rights. But, is land really immovable? For lawyers and legal scholars, this question must surely seem absurd, and they will answer it without hesitation in the affirmative. From a more sophisticated ontological perspective, however, matters are not so clear. For there is a range of types of immovable things whose treatment will shed light upon the partly fictional nature of the (positive) legal concept of immovability.

The standard classification of immovables stipulates four types:

- Immovables by nature, the paradigmatic examples of which are land parcels, edifices (including buildings) and plants adhering to the soil.
- Immovables by destination; here the best examples are agricultural machinery, animals associated with cultivation, and so on. These are all movable things that the law 'immobilizes' in order to account for the strict relationship of dependence in which these objects stand to other objects which are deemed immovables by nature.
- Immovables by the object to which they are applied; this category pertains to rights. This is a bold fiction of the law, for as Planiol points out: 'rights, being incorporeal are, strictly speaking neither movables nor immovables. They are not tangible. They take up no room' (Planiol 1930, 317). A classification of rights into movable and immovable can therefore be made only by attending to the object to which the right applies. If the right applies to an immovable thing, then the right is deemed immovable; if the right applies to a movable thing then the right is deemed movable.
- Immovables by declaration; finally, the category of immovables by declaration is the most fictional of all categories of immovable things, since here immovability is just a consequence of some individual's whim. Someone may, for example, simply declare some specific good to be immovable (for example, someone may declare an artwork in her own house to be immovable). There are stark differences from country to country in the way immovables by declaration are provided for and dealt with.

As can be clearly seen, the extent to which the immovability of an object depends on legal fictions varies considerably in the four cases mentioned. But it is hardly ever admitted that even in the case of land there is an element of fiction involved in its putatively immovable nature, and even in those rare cases where this element is indeed admitted, it is not further investigated. Planiol, for example, refers to that which is immovable by nature as follows:

> Strictly speaking, there is nothing which is absolutely immovable. Even the elements which compose the soil, rocks, sand, minerals, may be displaced. When a canal is dug, when lots are leveled it is the soil which is transported. In America, engineers have displaced large buildings without demolishing them. In Paris, the fountain du Palmier on the Place du Châtelet was set back in its entirety to permit the opening of the Boulevard de Sebastopol. But the law does not envisage the possibility of movement with the same rigor as mechanics. The law holds those things to be immovable [by nature] which are immovable in a durable and habitual manner and whose function is to be immovable,

even if they may be displaced, in some cases, by extraordinary means (Planiol 1930, 306).

Land moves, too, of course, with the movement of the earth (and a comprehensive analysis of land must take account of this fact if it is to do justice to the extension of property rights in land to the moon, or to distant planets, or even to entire sub-divisions of the cosmos). Even when we take account of the many fictions which it might be politically or economically or astronomically fruitful to allow, however, we must conclude that the initially plausible distinction between movables and immovables has only limited potential as the cornerstone of a rigorous analysis of landed property.

The Economics of Landed Property: What Can We Do With What We Own?

The economic effects of landed property are huge. A recent and comprehensive study (De Soto, 2000) highlights many of these effects. The central thesis of de Soto's book, which is entitled The Mystery of Capital, is that things do not amount to capital. Not even land amounts to capital. For as he points out: 'In Asia, Africa, the Middle East, and Latin America [...] most of the poor already possess the assets they need to make a success of capitalism' (Op. cit., 2000, 5). The problem is that 'they hold these resources in defective forms: houses built on land whose ownership rights are not adequately recorded, unincorporated businesses with undefined liability, industries located where investors cannot see them' (Op. cit., 2000, 5-6).

What De Soto seeks is an ontology of capital, along the same lines as the ontology of real estate that we sketch here. The crucial question that De Soto tries to answer is How do we transform things into capital? And of all the things that are so transformed, the most important, indeed the foundational one, is land. De Soto's book is provocatively wide-ranging and impressively researched; but its perspective is that of the economist, not that of the philosopher-ontologist, and it is precisely the latter that is needed if we are to make sense of the matters to which he draws attention.

De Soto, rightly, points out that those surprisingly abundant assets that the poor have in third world nations 'cannot readily be turned into capital, cannot be traded outside of narrow circles where people know and trust each other, cannot be used as collateral for a loan, and cannot be used as a share against an investment' (De Soto, 2000, 6). This is an extremely important point: the poor lack capital, but they do not necessarily lack assets (some of the poor could, of course, lack both, though the empirical evidence collected by De Soto strongly suggests that it is a lack of capital which is the problem). As a matter of fact, the difference between successful and unsuccessful nations, from the economic perspective, runs skew to the degree of development of their mechanism for turning stuff into capital.

Unfortunately, De Soto sometimes betrays the letter of his own thesis: he refers to these non-capitalized assets at times as 'non-capital' and at other times (more frequently) as 'dead capital' or as 'undercapitalized assets'. Of course, it might turn out that De Soto wishes that we understand these expressions as synonyms; nonetheless, it would be better if we had clear indications as to what exactly non-capital is, what capital is, what dead or dormant capital is (if such things exist) and how they all fit together within a single unified theory. And such a theory requires a further foundation in an ontology of real estate – for (as becomes clear through the course of De Soto's study) it is rights over land that are of paramount importance.

De Soto compares economically weak and underdeveloped nations to economically robust nations. The following holds only for the latter: 'every parcel of land, every building, every piece of equipment, or store of inventories is represented in a property document that is the visible sign of a vast hidden process that connects all these assets to the rest of the economy' (De Soto, 2000, 6). Though De Soto makes reference here to different types of objects and not only to land, it is clear that land is the most important of the objects which he seeks to investigate. It is not only the fact that 'the single most important source of funds for new businesses in the United States is a mortgage on the entrepreneur's house', and that mortgages are, in principle, applicable only to real estate (De Soto, 2000, 6). De Soto admits the primordial role of real estate also when, in explaining the comprehensive research agenda that led him and his associates to Egypt, Peru, Russia, Haiti, and the Philippines, he states that: 'To be more confident of our results, we focused our attention on the most tangible and detectable of assets: real estate' (De Soto, 2000, 30). What De Soto's research shows, in the end, is that a plausible and fruitful way to express the difference between developed and under-developed nations is the degree to which land is turned into real estate (and, of course, the degree to which that system which turns land into real estate then allows for further transactions with the fully capitalized parcels which result).

By 'raw land' in what follows we shall understand not real estate which is being under-utilized but rather physical land (of any sort) before it has become real estate. We can then affirm with De Soto that the cornerstone of the mechanism for turning raw land into real state – that is for turning stuff into capital – is a representational system made up of titles, deeds, registration documents, and so forth.

De Soto rightly insists that the representational system which is the basis of the formation of capital is not simply a collection of 'stand-ins for the assets': 'a formal property representation such as a title is not a reproduction of the house, like a photograph, but a representation of our concepts about the house' (De Soto, 2000, 50). An advantage of such representations is that, unlike physical assets, they are 'easily combined, divided, mobilized, and used to stimulate business deals' (De Soto, 2000, 56). But the most salient advantage of these representations is that they have the power to transform raw land into that multi-layered entity which is a parcel of real estate – or in other words to give rise to a plurality of ontologically distinguishable aspects of what is, from a geometrical point of view, identically the same piece of land. They thereby allow the fully capitalized assets to enjoy a multiple existence; namely, a physical existence, a legal existence, an economic existence, a political existence, and so forth. Compare De Soto's remarks on the differences between dwellings in economically developed and economically underdeveloped nations. In the latter, people's houses serve at best to protect them from the weather or from wild animals and criminals. In economically developed nations, in contrast, people's 'houses no longer merely keep the rain and cold out. Endowed with representational existence these houses can now lead a parallel life, doing economic things they could not have done before' (De Soto, 2000, 62-63).

It is clear then, that developing an accurate and efficient system of representation for land parcels, and of the transactions regarding these parcels, is a necessary condition for the functioning of capitalism in its developed form, and indeed of that transition to fully functioning capitalism which is economic development. Yet, as we shall see, the construction of such a system is not an easy task.

Collective Intentionality, Rules, and the Ontology of Property

Let us return to Rousseau's famous dictum quoted at the beginning of this essay. It is not only fencing off a plot of land that is important; important also is the fact that people believe that the person who fenced this plot of land is also the one who actually owns it. Collective intentionality is necessary for the existence of landed property. This is a crucial element of the ontology of property rights.

A recent and powerful attempt to apply ontological tools to the analysis of unorthodox entities like landed property is carried out by John Searle in his The Construction of Social Reality (Searle, 1995). Searle draws a distinction, first of all, between brute facts and institutional facts. Brute facts are those facts which exist independently of human conventions. Institutional facts are a sub-set of social facts; social facts are, simply those that depend on human conventions for their existence. The additional, special characteristic of institutional facts is that they involve the creation, extension or transfer of powers. Searle does not distinguish between rights and powers; as a matter of fact, whenever he speaks of powers in the realm of institutional facts he really means what are normally referred to as rights in our sense (for having a power, in the more usual sense, is typically a matter of brute facts, say, the power to invade your property). For the moment, nonetheless, we shall follow Searle in stating that the primitive term in the creation of social reality is power.

All institutional facts require collective intentionality. That certain rectangular pieces of paper count as money requires that there is a group of people who believe that they do so. (Which group of people is relevant for this purpose and how large it needs to be are difficult problems, which Searle does not discuss.) That Susan is French, that Manuel is Mexican are institutional facts, since nationalities, too, require collective intentionality. (That two plus two equals four, in contrast, is a brute fact, since it does not require collective intentionality.) That someone owns the shirt he is wearing requires collective intentionality, and so does the fact that someone owns a plot of land.

Searle has also put forth a now familiar distinction between what he calls regulative and constitutive rules. The former, as he puts it, merely regulate antecedently existing forms of behaviour. For example, the rules of polite table behaviour regulate eating, but eating itself exists independently of these rules. Some rules, on the other hand, do not merely regulate; they also create or define new forms of behaviour. The rules of chess create the very possibility of our engaging in the type of activity we call playing chess. The latter is just: acting in accordance with the given rules.

Constitutive rules, Searle tells us, 'always have the same logical form ... They are always of the logical form such-and-such counts as having the status so-and-so' (Searle 1999, pp. 123 f) An utterance of the form 'I promise ...' counts as putting oneself under a corresponding obligation. A given relationship between a person and a plot of land, counts as ownership. And as we see from these cases, the Y term in a constitutive rule characteristically marks something that has consequences in the form of rewards, penalties, or actions one is obliged to perform in the future.

When applying the X counts as Y formula we have to take into account whole systems of such rules. Acting in accordance with all or a sufficiently large subset of these and those rules by individuals of these and those sorts counts as conducting a legal trial according to Massachusetts law. The counts as formula can also be iterated so that whole systems of iterated structures (including the system we call property in land) can arise, systems which interact in multifariously spreading networks. Consider for example the way in which the marriage and inheritance systems have interacted with the landed property system in different cultures over time.

Searle's account of the way in which so much in human civilization rests in this way on systems of integrated and interleaved constitutive rules is certainly the most impressive theory of the ontology of social reality we currently have. But this account is also not without its problems, and the discussion of these problems sheds light on the ontology of landed property. For Searle's social ontology in its original form presupposes that – as in the case of President Clinton and Canterbury Cathedral and the money and driver's license in your pocket – the X terms at the bottom of the hierarchy are in every case parts of physical reality. When we examine the detailed workings of his theory, however, we discover that Searle is committed also to the existence of what we might call 'free-standing Y terms', or in other words to entities which do not coincide ontologically with any part of physical reality. One important class of such entities is illustrated by what we loosely think of as the money in our bank accounts as this is recorded in the bank's computers. In The Construction of Social Reality we find the following passage:

> all sorts of things can be money, but there has to be some physical realization, some brute fact – even if it is only a bit of paper or a blip on a computer disk – on which we can impose our institutional form of status function. Thus there are no institutional facts without brute facts (Searle 1995, 56).

On closer inspection, however, it becomes clear that blips in computers do not really count as money and nor can we use such blips as a medium of exchange. Rather, as Searle has subsequently acknowledged, blips in computers are merely representations of money, and as he points out, it would be a 'fascinating project to work out the role of these different sorts of representations of institutional facts' (Searle 2002).

Searle here recognizes a new dimension in the scaffolding of institutional reality, the dimension of representations. As the blips in the bank's computers merely represent money, so the deeds to your property merely record or register the existence of your property right. The deed is not identical with your property right and nor does it count as your property right. An IOU note, similarly, records the existence of a debt; it does not count as the debt. It is an error to run together records pertaining to the existence of free-standing Y terms with those free-standing Y terms themselves.

As the case of money shows, some social objects have an intermittent and what we might think of as a merely generic realization. Others, such as corporations or universities, have a physical realization that is partial and also scattered (and also such as to involve a certain turnover of parts). Yet others, such as debts, may have no physical realization at all; they exist only because they are reflected in records or representations (including mental representations). A full-dress ontology of social reality must address all of the different types of cases mentioned, from Y terms which are fully identical with determinate parts of physical reality to Y terms which coincide with no determinate parts of physical reality at all.

Free-standing Y terms, as might have been predicted, are especially prominent in the higher reaches of institutional reality, and especially in the domain of economic phenomena, where we often take advantage of their abstract status in order to manipulate them in quasi-mathematical ways. Thus we pool and securitize loans, we depreciate and collateralize and amortize assets, we consolidate and apportion debts, we annualize savings – and these examples, along with the already mentioned example of the money existing (somehow) in our banks' computers, make it clear that the realm of free-standing Y terms must be of great consequence for any theory of institutional reality.

That this is so is made abundantly clear not least by De Soto's work – which was indeed in part inspired by The Construction of Social Reality and which also goes some way towards realizing Searle's 'fascinating project' of working out the role of the different sorts of representations of institutional facts. As De Soto shows, it is the 'invisible infrastructure of asset management' upon which the astonishing fecundity of Western capitalism rests, and this invisible infrastructure consists precisely of representations, for example of the property records and titles which capture what is economically meaningful about the corresponding assets – representations which in some cases serve to determine the nature and extent of the assets themselves. (See Smith and Zaibert, 2001).

Capital itself, in De Soto's eyes, belongs precisely to the family of those free-standing Y terms which exist in virtue of our representations:

> Capital is born by representing in writing – in a title, a security, a contract, and other such records – the most economically and socially useful qualities [associated with a given asset]. The moment you focus your attention on the title of a house, for example, and not on the house itself, you have automatically stepped from the material world into the conceptual universe where capital lives (De Soto 2002, pp. 49 ff.).

As those who live in underdeveloped regions of the world well know, it is not physical dwellings which serve as security in credit transactions, but rather the equity that is associated therewith. The latter certainly depends for its existence upon the underlying physical object; but there is no part of physical reality which counts as the equity in your house. Already the term 'negative equity' should draw our attention to the special nature of this phenomenon. Equity is tied to time, to history, and to a certain portion of physical reality; yet it is at the same time something abstract, something that exists only insofar as it is represented in a legal record or title in such a way that it can be used to provide security to lenders in the form of liens, mortgages, easements, or other covenants in ways which give rise to new types of institutions such as title and property insurance, mortgage securitization, bankruptcy liquidation, and so forth.

The Uniqueness of Landed Property

Landed property in general is nestled in a much more complicated system of constitutive rules, and it requires more variegated forms of collective intentionality than do other forms of property. There is a sense in which the existence of any right whatsoever requires collective intentionality. Unless one believes in the existence of some form of natural law which would imply the existence of rights independent of any human conventions, any right requires for its existence that people believe that it is indeed a right. In the simplest case, someone might have property rights over the shirt he is wearing. The only aspect of this situation that requires collective intentionality is that relating to the institutional fact: this person owns this shirt. In the case of property in land, however, collective intentionality is required not only at the level of the person owning the land but also with respect to the existence of the very plot of land itself. For here it is not only the property right itself that requires collective intentionality, but also the object over which the right falls.

We suspect that this explains Rousseau's characteristically malicious suggestion that the people who would believe that the plot of land is indeed the property of the person who fenced it off are simpletons – people who have been duped. It would have been less easy for Rousseau to make this same point in respect to, say, those of his fellows who believed that Rousseau himself was the owner of the shirt on his back. This is because, in relation to the ownership of the shirt, there is one level only that is subject to collective intentionality. In relation to the plot of land, in contrast, it is not only in the existence of the right of property that we have to believe, but also in the existence of the very object over which the property right falls – an object which is supposed to be somehow created by the very act of fencing off.

Some political discussions regarding property rights do indeed recognize the distinction between landed and other forms of property. For example Henry George called for the institution of a 'single tax' on land, on the grounds that one cannot legitimately own naturally occurring resources, but can only have rights to the value one adds through one's own work – a proposal that has been endorsed in our own day by Hillel Steiner (1994). And as Richard Pipes reminds us, John Stuart Mill questioned whether land should be treated as merely one particular form of property, on the grounds, first, that no one had made it, and second, that whereas in creating movable wealth one did not deprive one's fellowmen of an opportunity to do likewise, in appropriating land one excludes others (Pipes 1999, 57).

The contrast drawn by George is far from being absolute, however. Thus it may take work (and the adoption of considerable risks) to discover natural resources such as gold, and land, and if all natural resources were to count as common property, then much of this work (and risk) would not be forthcoming. Mill's criterion of excludability is on the right track. But it captures only part of what is, from the ontological point of view, a much more complex phenomenon.

What is a Property Right?

Property rights are complex sets of other rights, and excludability is only one of the many rights in the bundle, and land is different from other forms of property also for reasons which have to do with features of this complex set. Property is often conceived, à la Hohfeld (1919), after the model of a bundle of sticks. Each stick in the bundle signifies a particular right or power: a right to use, a right to possess, to sub-divide, to rent, to build upon, to enjoy the usufruct from, and so on. An owner can, in certain cases, sell or give away specific rights, or see these rights removed, divided, or amended by the force of others. Our practical dealings with landed property in cases where the sticks have dwindled or been transformed in this fashion can be a very complex matter. It is important to point out, however, that the absolute property right itself is in no way affected by this dwindling of the rights (or powers) that make up the property right. This means that Hohfeld's 'bundle' analogy is in fact not quite correct, though we shall find it useful to employ his terminology nonetheless. As Reinach has eloquently put it:

> If property were a sum or unity of rights, it would be reduced by the alienation of one of these rights, for a sum necessarily disappears with the disappearance of all its parts. But we see that a thing continues to belong to a person in exactly the same sense, however many rights he may want to alienate; it makes no sense at all to speak of a more or less with respect to belonging. The nuda proprietas in no way means that the owning 'springs back to life' once the rights transferred to other persons have been extinguished; the thing rather belongs to the owner in the interval in exactly the same sense as before and after ... This is the essential necessity which underlies the so-called 'elasticity' or 'residuarity' of property and which can hardly be reasonably considered as an 'invention' of the positive law (Reinach, 1987, 56).

Each of the sticks that make up the property right can, in principle at least, be the object of negotiations independently of the remaining sticks in the bundle, and whatever the outcome of such negotiations the property right – the absolute relation of belonging – remains ontologically speaking intact. Someone can give away some of the sticks without giving away his property over the thing in question. Thus it is not uncommon to see cases in which someone has given away (or has had taken away) virtually all the sticks in the bundle (in the case, for example, of the possession of his land by squatters); but even then his residual property right over the thing itself remains.

The bundle of property rights in land has first of all the elastic or residual character that has been referred to already above. Such elasticity is manifested to some degree in other spheres, for example in the car rental or equipment leasing markets. But it still seems odd to suppose that someone might give away the right to use a washing machine or toothbrush for long periods of time while retaining title to the goods in question. In most such cases it seems that, when someone gives away a specific stick from the bundle, then he is actually giving away the full right of property over the object in question.

Two interconnected reasons explain why it is especially in the case of landed property that this residual character is essential. First, some types of negotiations relating to the sticks in the bundle make practical sense only in relation to landed property. Although the owner of, say, a painting, or a car, strictly speaking has the right to subdivide it, it seems unlikely that he will ever seek to exercise this right.

Second, it is primarily in relation to landed property that the mentioned maneuvers (subdividing, commercializing the fruits of, etc.) are commonly carried out, precisely because there are here more sticks in the bundle, and they are more varied and complex than in relation to other types of property. Leasing, time-sharing, owning shares in a social club, borrowing, sub-dividing, using as collateral are examples which illustrate just some of the possibilities here. And because of the central economic importance of land as the presupposition of all other human activity, it is only in the case of landed property that correspondingly complex legal institutions have grown up in reflection of the different dimensions of rights involved.

Consider, for example, my property right over my watch: it is easy to see that the bundle of sticks which comprises this property right can only be altered with difficulty – and even then still only partially. We cannot, after all, meaningfully talk about subdividing, or building upon a watch, or harvesting the usufruct therefrom. What purpose could be served by giving away the possession or the use of the watch while maintaining ownership over it? The age-old aphorism 'possession is nine tenths of the law' is, under this light, exactly right. While ownership and possession are closely related phenomena, the relationship between them is much closer in the case of movables than in the case of immovables.

A further important reason for the differences between landed property and other types of property turns on the special geographic dimension of the objects of property rights in land. As we have seen, the idea of a parcel of land is in greater need of ontological clarification than is, say, that of a watch or a lawnmower. A parcel of land, we can now say, has fiat boundaries, and this means: it needs to have its boundaries provided for by some human institutions. A full-blown ontological analysis of real estate must thus provide an account not only of the make-up of the bundle of sticks which comprises a property right in general, but also of the accompanying institutions for example of boundary maintenance and title and cadastral registration. It must also provide an account of the interplay between these dimensions – and this in such a way as to do justice also to the differences between different human cultures. The analysis in question must accordingly have at least the following components, each one of which will be seen to have been at work in the arguments above:

- When someone owns a parcel of real estate, then there is a certain portion of the surface of the earth to which he is related.
- This portion of land must have the character of an enduring object which – at least when considered on the scale of human events – endures permanently.
- This portion of land must have definite, known (or at least knowable) boundaries.
- The portion of land must be such that the owner, and in principle others, may gain (legal and physical) access.
- The portion of land must be knowable. Investors and others must know where it is situated.
- Real estate gives rise to neighbors. There are no neighbors where there is raw land, simply because they are no boundaries in raw land. Even the so-called bona fide boundaries – those obvious discontinuities on the surface of the earth, such as coastlines, mountain ranges, rivers, etc., are not boundaries in the sense which pertains to the ontology of real estate – until someone considers them to be so.
- Parcels of real estate have different conditions of identity than do raw land. I might exchange all the soil in my land in New York for the soil in your land in Delaware, yet I would still be the owner of real estate in New York and you in Delaware.
- A parcel of real estate is multi-layered in the sense that there are ontologically distinguishable aspects of what is, from a geometrical point of view, identically the same piece of land. There are layers of geology, of archeology, of history, of ecology, of rights of way, and so on, as well as layers of insurance, equity, and economic value; and the state can own (or have property rights in) some or all of these layers even in those circumstances where a private person is the ostensible owner of the plot of land simply conceived.
- A parcel of real estate is a three-dimensional solid which includes regions above and below the surface of the earth itself. As an owner of a parcel of real estate I must for example have the right to prohibit my neighbor from building a structure that would invade the space above my land. This feature illustrates most clearly the institutional (fiat) character of real estate. For even in regard to pure geometry, the specification of the height and depth of the relevant three-dimensional solid differs from culture to culture. In the United States, for example, the owner of a given parcel in fact (and in law) owns a cone-shaped region of space projecting from the center of the earth and reaching upwards (roughly) as far as the ear can hear. In other places these determinations are effected in different ways. One of the specific prerogatives which the state has

in Latin America is that it owns the whole of the subsoil in the country, no matter who owns the surface of the land.

- The boundaries of a land parcel are affected by a factor which we might call crispable vagueness – that is by a vagueness that can, where necessary for practical reasons, be alleviated by institutional fiat or by negotiation. (Smith, 2001) If someone owns a land-parcel in Venezuela, and finds gold some few inches below the ground, this gold becomes the property of the state. Of course, this presents the state with the problem of determining how to fix the boundary between the surface and the subsoil. It seems odd, to say the least, that a hand-made hole of merely a few inches constitutes a penetration in the state's exclusive property. Note that the problem faced by even developed institutions of property law in providing a clear demarcation of such a boundary is analogous to the problem of drawing a line between, say, territorial and extraterritorial waters. Fiat crisping will occur only where it is of practical importance. Cadastral and title registration, for example, is much more precise and reliable in countries, such as Switzerland or Austria or Holland, were land is scarce, than it is in the US or Australia or (presumably) Siberia.

Appendix: Apriorism, Realism, and the Ontology of Landed Property

In The A Priori Foundations of the Civil Law, Reinach sought to attack the view that the concepts and structures of the civil law were created by the civil law, that is, that they were merely the reflections of laws as created human institutions. Reinach, in contrast, sought to show that 'the positive law finds the legal concepts which enter into it; in absolutely no way does it produce them [emphasis in the original]' (Reinach, 1983, 4). Thus, Reinach attacks precisely the sort of view that Searle puts forth. Reinach further tells us that specifically legal structures 'have a being of their own just as much as numbers, trees, or houses' and 'that this being is independent of its being grasped by men' (Reinach, 1983, 4). There are true propositions in the realm of the law, he held, and these propositions are true independently of anyone knowing that they are true and of anyone deciding to create the concepts to which the propositions refer.

Reinach's thus embraces a doctrine of apriorism regarding the basic building blocks of the legal realm, a doctrine which he takes as providing a bulwark against legal positivism, legal relativism and related positions. Reinach is thus not merely a realist about legal institutions. A realist in regard to a given domain holds that there are facts pertaining to that domain which obtain independently of whether or not they are recognized as obtaining. Reinach goes further in embracing legal apriorism: he holds, in other words, that there is a special way in which we come to know these recognition-transcendent facts.

'If there are legal entities and structures which in this way exist in themselves', Reinach points out, 'then a new realm opens up here for philosophy. Insofar as philosophy is ontology of the a priori theory of objects, then it has to do with the analysis of all possible kinds of object as such' (Reinach, 1983, 6). True to this goal, most of Reinach's book is devoted to an analysis of basic legal concepts such as claim, right, obligation, promise, property, and so on. In light of Reinach's analysis, moreover, law resembles certain other disciplines: like 'pure mathematics and pure natural science there is also a pure science of law' (Reinach, 1983, 6). Already in 1869 Ernest Beling, Reinach's teacher, had attempted an aprioristic analysis of the criminal law in his Die Lehre Vom Verbrechen. Carl Menger attempted to deploy apriorism as a basis for the science of economics and in this he represents faithfully the spirit of the so-called 'Austrian School', which he founded (see Menger 1871).

The connection between Reinach's apriorism and the ontology of landed property can now be explained as follows. Given the multi-layered ontology of landed property, Searle's simple ontology based on collective intentionality cannot do the work. Someone owning a given plot of land is not, under normal circumstances, affected by the collective beliefs of any group, even though those beliefs were perhaps necessary to set up the relevant system of landed property in the first place. He may just own the land, independently of the beliefs of those around him. While collective intentionality is thus perhaps crucial for the creation of institutional reality, and also for the resolution of disputes concerning this reality, it is not so important for the continued existence of this reality in the normal case. Moreover, there are important aspects of the phenomenon of real estate which are not the result of human agreements of any sort. Rather, they are part and parcel of the underlying structure of real estate as such, a structure which is intelligible to beings like ourselves, not because we have created it but because, like the structures of promising, claim, obligation, debt, and so forth, and also like the structures of circle, square, triangle, hypotenuse, it is there waiting to be discovered. In Reinach's words, there are certain basic legal entities and structures which exist independently of the positive law, though they are presupposed and used by it. Thus the analysis of them, the purely immanent, intuitive clarification of their essence, can be of importance for positive-legal discipline. The laws, too, which are grounded in their essence, play a much greater role within the positive law than one might suspect. One knows how often in jurisprudence principles are spoken of which, without being written law, are 'self-evident', or 'follow from the nature of things' to mention only a few of these expressions. In most cases it is not a matter of principles whose practical usefulness or whose justice is fully evident, but rather the essential structures investigated by the apriori theory of right. They are really principles which follow from the 'nature' or 'essence' of the concepts in question (Reinach, 1983, 6-7).

That an obligation ceases to exist after it has been discharged is a principle that has nothing to do with any agreement between men; the validity of this principle does not presuppose intentional states of any kind. If someone understands the concept of obligation, he will ipso facto realize that it would make no sense to suggest that someone under an obligation to do X remains obligated after doing X. Similarly, and more concretely, that real estate must have boundaries, or that it must give rise to neighbors, or, in general, that the ontology of real estate must do justice to the characteristics listed above, is not an empirical discovery (or the product of some convention) but rather a matter of the intelligible structure of the domain in question. It should be clear how Reinach's approach differs from that of Searle. While there is no doubt that Searle provides a valuable analysis of the ontological structures underlying many institutional phenomena, his framework allows too much to be the result of fiat and convention. And in fulfilling the task of the ontology of real estate we need to take into account not only those dimensions of the realm of landed property which are conventional in nature, but also those dimensions which are prior to all conventions – and which thus make these conventions possible.

References

Beling, Ernest. (1964), *Die Lehre Vom Verbrechen*, Tübingen: Scientia Verlag Aalen.

Bentham, Jeremy. (1958), 'Principles of the Civil Code', in John Bowring, (ed.), *The Works of Jeremy Bentham*, Vol. 1, New York: Russell and Russell.

Bittner, Steffen, Wolff, Annette von, Frank, Andrew U. (2000), 'The Structure of Reality in a Cadastre', in Berit Brogaard (ed.), *Rationality and Irrationality* (Papers of the 23rd International Wittgenstein Symposium), Kirchberg am Wechsel: Austrian Ludwig Wittgenstein Society, 88–96.

De Soto, Hernando. (2000), *The Mystery of Capital*, New York: Basic Books.

Hohfeld, Wesley. (1913), 'Some Fundamental Legal Conceptions as Applied in Judicial Reasoning' *Yale Law Review* 23.

Mark, David M., Smith, Barry and Tversky, Barbara. (1999), 'Ontology and Geographic Objects: An Empirical Study of Cognitive Categorization', in C. Freksa and David M. Mark (eds.), *Spatial Information Theory. Cognitive and Computational Foundations of Geographic Information Science* (Springer Lecture Notes in Computer Science 1661), 283–298.

Menger, Carl. (1871), *Grundsätze der Volkswirtschaftlehre*, Vienna: Braumüller.

Munzer, Stephen R. (1995), *A Theory of Property*, Cambridge: Cambridge University Press.

Munzer, Stephen R. (1994) An Uneasy Case Against Property Rights in Body Parts, *Social Philosophy and Policy*, 11: 2, 259-86.

Nozick, Robert. (1974), *Anarchy, State and Utopia*, New York: Basic Books.

Pipes, Richard. *Property and Freedom*, New York: Vintage, 1999.

Planiol, Marcel. (1939), *Treatise on the Civil Law*, Lousiana State Law Institute translation).

Reinach, Adolf. (1983), 'The A Priori Foundations of the Civil Law' *Aletheia* (3): 1-143.

Rousseau, Jean-Jacques. (1992), *Discourse on the Origins of Inequality*, Indianapolis: Hackett.

Ryan Alan. (1994) Self Ownership, Autonomy, and Property Rights, *Social Philosophy and Policy*, 11: 2, 241-258.

Searle, John R. (1995), *The Construction of Social Reality*, New York: Free Press.

Smith, Barry, (1990), 'The Question of Apriorism', *Austrian Economics Newsletter*: 1-5.

Smith, Barry (1992), 'An Essay on Material Necessity', in P. Hanson and B. Hunter, eds., *Return of the A Priori* (Canadian Journal of Philosophy, Supplementary Volume 18).

Smith, Barry. (2001), 'Fiat Objects', *Topoi*, 20: 2, September 2001, 131–148.

Smith, Barry (forthcoming), 'Ontology', in Luciano Floridi (ed.), *Blackwell Guide to Philosophy, Information and Computers*, Oxford: Blackwell.

Smith, Barry and Brogaard, Berit (2002), 'Sixteen Days', *The Journal of Medicine and Philosophy*.

Smith, Barry and Mark, David M. 1999 'Ontology with Human Subjects Testing: An Empirical Investigation of Geographic Categories', *American Journal of Economics and Sociology*, 58: 2, 245–272.

Smith, Barry and Mark, David M. (2001), 'Geographic Categories: An Ontological Investigation', *International Journal of Geographic Information Science*, forthcoming.

Smith, Barry and Zaibert, Leo. 'The Metaphysics of Real Estate', *Topoi* 20 (2001): 161-172.

Spector, Mary B. (1986), 'Vertical and Horizontal Aspects of Takings Jurisprudence: Is Airspace Property?', *Cardozo Law Review* 7: 489-518.

Steiner, Hillel. (1994), *An Essay on Rights*, Oxford: Blackwell.

Stubkjær, Erik (2001), 'Spatial, Socio-Economic Units and Societal Needs – Danish Experiences in a Theoretical Context', in A. Frank et al. (eds.), *The Life and Motion of Socio-Economic Units* (GISDATA 8), London: Taylor and Francis, 265–280.

Thomasson, Amie L. (1999), *Fiction and Metaphysics*, Cambridge: Cambridge University Press.

Zaibert, Leo. (1999), 'Real Estate as an Institutional Fact: A Philosophy of Everyday Objects', *American Journal of Sociology and Economics*, 58: 2, 273-284.

PART II
REQUIREMENTS AND NATIONAL PERSPECTIVES

Chapter 4

Purchase of Real Property in Finland

Kauko Viitanen

Abstract

In this article, the process of purchasing real property in Finland is treated. The owner of a real property may sell it freely by him/herself but normally a commission with a real estate broker for selling the property is made. A written sales contract must be signed by the seller, the buyer, and by a public purchase witness. The buyer must take care of the registration of the deed (apply title) and pay the transfer tax (four per cent). The commission fee to the broker (about 4 per cent + VAT) is normally paid by the seller.

Introduction

The aim of this article is to provide a clear picture of the main phases and actors in the process of purchasing real property in Finland in the most common situations. Consequently, the study concentrates on the purchase of a subdivided plot with a built-up one family house in a local detailed plan area in accordance with the plan (real property in a detail plan area). In addition, it is assumed that the owner (seller) is a physical person with a title to the real property and that he/she is the only owner and user of the property. The buyer will also be a physical person who buys the whole property. There is no need for a change of the plan.

These assumptions are made because this article is a part of an EU-project which aims to compare and provide initiatives for making the present real property transaction procedure more effective and easy to understand, i.e. transparent. In the starting phase, it is easiest to find the basic elements of transaction processes in the normal house transaction of families. These kinds of transactions are continuously performed in countries with private property ownership. After an examination of the basic transactions, the study can more easily move to more complicated situations.

The article first explains different conditions for and around purchases. After that, a typical purchase procedure is presented.

Real property in Finland

The Finnish territory is divided into real properties and other register units. A real property is a unit of ownership which must be registered in the cadastre as a real property (Real Property Formation Act 2 §). There are nine types of real property units. This article concerns two of the most common types of real properties, the 'estate' (tila) and 'site' (tontti). They are normally under private ownership. The other seven types are mainly public: land and water areas with special restrictions. Other register units consist of state roads and common areas (indirect ownership).

A real property unit is an item of ownership, use, conveyance, and mortgage. A real property may be owned by private persons or legal persons, i.e. by companies and other entities, or by the state and local municipalities. It may consist of several separate parcels of land or water. Buildings belonging to the same owner as the land they are located on constitute part of the real property (fixtures). If the buildings have a different owner to that of the land, e.g. based on leasehold, the ownership of buildings is regarded as personal property. In a real property unit, other types of rights may additionally be granted (Viitanen, et al., 1997).

The proprietary rights are entered two-dimensionally in the cadastre and the land register. The mortgages may concern only the real properties. The perpetual easements and other usufructs are entered in the cadastre under the encumbered and, if possible, the justifiable real estate. Temporary restrictions (rights) of usufructs based on an agreement are entered in the land register (Land Code).

Land use planning

General responsibility for planning and the development of land use is vested in the Ministry of the Environment. This includes the direction and supervision of physical planning and building, as well as housing policy and environmental protection. Guidance, supervision and development of the physical structure of the environment are normally regulated by the Land Use and Building Act (LBA). This act and its complementary statutes include the principal regulations on planning and building.

There are three different plan levels: the regional plan, the local master plan, and the local detailed plan. Besides these plans, there are additionally local building ordinances in the municipalities. These compulsory regulations include detailed guidelines and rules for building in the municipality. All forms of planning are normally legally binding after being approved by a municipal council. No plan as such implies the right to build. A building permit from the municipal building committee is always needed.

In city areas, the structure of real properties is defined in the local detailed plan, i.e. the real properties must be formed to be fitted to the plan. Consequently, a real property in a detailed plan area normally consists of only one parcel. When the property has been formed and built up according to the plan, the land use may only be changed after changing the detailed plan, which is normally not an easy process. In other words, the owner's rights in a built-up site are well protected.[1]

Ownership of real property

Private persons, companies, and other economical or juridical entities may own a real property with no general limitations. Earlier, there existed restrictions for foreigners and foreign companies in possessing and owning real property but the last restrictions were abolished at the beginning of 2000. Today, there are no general limitations, except in the Åland (Ahvenanmaa) Province (archipelago), where the right of domicile on the island is required even for the Finns (Viitanen, et al., 2002).

The ownership of real property must be registered in the public Land Register, which is kept by the local courts. A title is given to the owner as a document of registration. The granting of title does not create the right of ownership. It is only a public declaration of the right of ownership, the owner, the property in question, and the purchase process. However, the register meets officially the requirements of public reliability (Viitanen, et al., 1997, p.46).

Legal terms for purchase of a real property

The transaction of real property is regulated in the Land Code. Buying real property is an official legal act, which must be performed in a specified form. This includes a written document (sales contract) with the principal terms, the signatures of the contracting parties, and attestation by a public purchase witness. The signed contract (i.e. the ownership) must be registered in the local court. Before the deal is closed, the seller is obliged to give all relevant information to the buyer but the buyer shall check the property.

Sales contract

The sales contract must be a written document signed by the seller and the buyer or by their representatives, and attested by a public purchase witness when the signatories of the deed of sale (seller and buyer) are present. When a married private person is selling a property used as a family home, an approval from the

---

[1] The right of ownership may be expropriated under special conditions but then the owner is entitled to full compensation (Constitution 15 §). However, in certain cases there are exceptions of the rule, e.g. the landowner is obliged to give up the area needed for streets to the municipality without compensation according to the first local detailed plan (LBA 104-105 §§). Currently, the area transferred without compensation shall not exceed 20 per cent of the total land owned by the landowner in the plan area in question, or shall not be larger than the building volume permitted for the land remaining in his/her ownership. However, this does not apply in the present case, where it is assumed that the real property is already implemented according to a detailed plan.

husband or wife is normally required, although the other partner may not own a share of the property. A transaction is not binding if not made according to the law (Land Code).

According to the Land Code (2:1), a sales contract must include at least (obligatory terms):

- a clear declaration to transfer the property from the seller (transferor) to the buyer (transferee)
- a specification of the real property concerned by the sale
- identification of the parties of the sale, i.e. the seller and the buyer
- the purchase price or other compensation.[2]

Normally, the contract additionally includes at least the following terms (Kasso, 2001, pp. 345-349):

- terms and time of payment
- time of the transfer of the possession of the real property (change of the ownership)[3]
- mortgages, if any (mortgaged pledges are normally used as a security)
- responsibility for different charges
- limitations of liability for the building
- other encumbrances.

Suspensive and/or dissolving provisions must be included in the contract in order for it to be effective (Land Code 2:2). They can be valid for five years. There are additionally certain types of provisions that are not legally binding although they have been included in the contract (Land Code 2:10).[4]

A minimum of three copies of the sales contract must be made, one for each party and one for the public purchase witness.[5]

Duties of the parties

The seller has duties to the buyer. All information provided to the buyer by the seller must be true and sufficient, and all information that may have an influence on the buyer's decision must be given. Furthermore, the buyer is additionally obliged to investigate the property thoroughly, and cannot afterwards claim for

---

[2] The buyer has no right to charge more than the price revealed in the sales contract (Land Code 2:1).

[3] If not otherwise agreed, the time of the transfer of the possession is the date of the contract (Land Code 2:12).

[4] According to general principles, restrictions on buyer's right of full ownership are prohibited.

[5] More precisely about the purchase of real property see e.g. Kasso (2001, pp. 332-360).

such things that he/she should have noticed before the decision (Land Code 2:17-22).

Various problems may arise subsequent to the purchase, for example if the contract includes false information about the land or the building, or if the building is in worse condition than that stipulated in the contract, or if the building has faults not visible during the inspection. Environmental matters have also become increasingly important. In recent years, it has been normal practice to permit specialists to investigate the building and the soil before the purchase.[6]

If the sold property is not in the condition that the buyer has expected and the seller has described, the buyer may recover the payment equal to the fault, or in more serious cases, the buyer may have the right to cancel the trade. The liabilities of the seller may be limited in the purchase agreement but the limitations must be determined in detail. Normally the term guarantee is five years from the purchase. Claims based on incorrectness must be presented within the designated period, which is prescribed by the law, and is normally a maximum of five years from the right of possession (Land Code 2:25).

Incorrectness related to the seller's responsibility for real property can be (Land Code 2:17-19):

- quality error, whereby the quality of the property does not correspond to that agreed
- legal error, whereby there is obscurity in the ownership or other rights related to the real property
- utility error, whereby the right to use the property is restricted without the knowledge of the buyer.

Public purchase witness

To be legal, the sales contract of real property must be attested by a public purchase witness (Land Code 2:1). The purchase witness acts as a witness to the purchase contract, and must check that the deed complies with legal formalities, i.e. that the contract includes the obligatory terms (see the list before), but the witness is not responsible for the content of the contract. The witness must further check the identity of the purchase parties and sign the contract when the other signers of the deed of sale are present, i.e. the signing is done simultaneously. At the request of the buyer or seller, the witness must check the information of the real property in the Land Register and in the Cadastre. After the deal is signed the purchase witness instructs to the buyer as to how to register the deed (obtain the title). In seven days, the public purchase witness must report the purchase to the District Survey Office, which registers it in the public purchase price register, and the municipality concerned. In addition, the tax office will be informed. The purchase witness must retain one signed document in his/her archive (Viitanen, et al., 2002).

---

[6] This is particularly important when the land earlier has been in an industrial use.

According to the Degree on Public Purchase Witness (958/1996), public purchase witnesses may include:

- such civil servants as a census officer, public notary, police chief, cadastral engineer (land surveyor) and
- persons nominated by the local court.

There is a register of public purchase witnesses, each having a unique identification number.

The fee for attestation is regulated. In 2000, the fee was approximately €70 per purchase. In addition, the witness is entitled to a travel allowance. (Kasso, 2001, pp. 336-338). The cost is normally shared equally between the seller and the buyer.

Registration of the deed and mortgaging the property

According to the Land Code (Chapters 11-12), the buyer shall apply for a title from the local court within six months of signing the contract. The application must include, e.g.:

- certificate of the seller's ownership (seller's title)
- the original sales contract and a copy of it
- certificate that the transaction tax (4 per cent of the purchase price) has been paid (normally by the buyer)
- certificate of change of ownership (normally expressed in the deed).[7]

The court examines the legality of the deed. If all particulars are clear, the court registers the deed in the public Land Register, transfers the title to the new owner (buyer), and returns the original sales contract to the buyer. The buyer shall pay the registration fee.[8]

The mortgage is applied in a similar manner quite similarly to the title. Mortgaging is an official process and is carried out in the local court, conventionally by the property owner. Normally, the buyer (owner) applies for a mortgage when applying for the title. The court will give a special mortgage letter as a document of the mortgage. This letter may be pledged, and through this process the creditor gets a lien for the real property (mortgaged loan). The mortgages are registered in the Land Register. There are no longer such tax costs as stamp duty for mortgages or loans, except small payments for the documents (Viitanen, et al., 2002).

---

[7] This may, for example, have been stipulated in the contract to depend on the date of paying the sales price.

[8] It is strongly recommended that the title be applied for directly after signing the sales contract because, until the registration of the application, the seller can apply for a new mortgage for the 'old' property (e.g. Kasso, 2001, p.356).

Normally, registration and mortgaging is taken care of by the buyer's creditor (a bank) as the buyer's representative. If the real property is mortgaged before the purchase, the mortgage letters should be given to the buyer after signing the sales contract and after the buyer has paid the purchase price to the seller.

Pre-emption of the municipality

According to the Pre-emption Act, municipalities have the right of pre-emption in certain cases of transactions. If the municipality decides to use its right of pre-emption, the buyer is replaced by the municipality in the original sales contract on the terms mentioned. However, the right of pre-emption is not directed towards small properties (i.e. not to properties under 5,000 sq.m., or under 3,000 sq.m. in the capital area); therefore, the Act will normally not be in effect in transactions of single-family houses. In practice, the right is relatively seldom used even in other cases.

Taxation

There are many different taxes connected to the ownership and transaction of real property. The real estate tax paid to the municipality where the property is located varies between 0.22 per cent and 0.5 per cent of the assessed value of residential properties.[9] In addition, there is a wealth tax, 1 per cent of the wealth over €185,000. For sales profit and rental revenues, there is an income tax of 29 per cent (year 2002) and a municipal tax, normally 15 to 20 per cent of the net income. The value added tax (VAT) payable on goods and services (e.g. construction, consultation, property asset management, and brokerage) is 22 per cent (year 2002). The most important tax in real property transaction is undoubtedly the transfer tax.

In the transaction of a real property, the buyer is obliged to pay a transfer tax before assuming title to the property (Land Code 2:15; Kasso, 2001, p.354). The transfer tax for real property is four per cent of the sales price[10] (or value of the property). The tax must be paid within six months of the purchase. If the registering of the deed is delayed, the transfer tax will be raised 20 to 100 per cent of the amount of the original tax.[11]

---

[9] The registered owner is responsible for the real estate tax in the beginning of the year, as is the new owner if the property has been sold during the year. Normally, the tax is paid in the autumn.

[10] The average price of a subdivided plot with a built-up one-family house in a local detailed plan area was €100,000 nationally, and €165,000 in the county of Uusimaa in the southern Finland (includes the capital city) in 2000. The number of sales was 8,321 nationally and 1,360 in the county of Uusimaa (NLS, 2001, p.37).

[11] People who have permanent residence in Finland and Finnish companies are generally taxable for their incomes. Other companies or individuals without a residence in Finland must pay tax on incomes they earn in Finland. To avoid double taxation, Finland has tax

Real estate agencies (brokerage)

Approximately half of all residential property sales are made with the help of real estate agencies or brokers.[12] Most often, the agencies are specialized in either residential or commercial property. The agencies may be limited companies with hundreds of persons or small entities involving only one entrepreneur. All agencies must be registered with the local provincial administrative board. The registered brokers may use the letters 'LKV' (real estate broker). To be registered, the agency must have a manager in charge who has passed a special examination for brokers. The examination is organised by the Central Chamber of Commerce. The registered agencies must obtain insurance against liability risks (Kasso, 2001, pp. 461-510).

The real estate agencies act as an intermediary between the seller and the buyer. Although the broker has an assignment with one party, usually the seller, the broker is obliged according to the law to protect the other party's, that is the buyer's, interests. The principal task for the brokers is to bring the seller and the buyer together. When this is done, the broker normally makes the documents for the transaction. The broker additionally has a responsibility to find out all information, necessary for the assignment, and the broker may be responsible for negligence. The obligations of the agency in the residential sector are specified in legislation.

Normally, agencies invoice a commission only of successful transactions. If the buyer cannot be found (when the client of the agency is the seller), no commission to the broker will normally be paid, expect certain costs mentioned in the assignment. Normally the commission is a percentage of the purchase price. In residential properties, the commission varies between two per cent and five per cent of the purchase price, excluding VAT. The commissions for the broker are subject to value added tax, which means that the VAT of 22 per cent (year 2002) of the commission must be paid. As the most common commission is four per cent of the purchase price, the VAT increases the payment to 4,88 per cent.

Real estate valuation

A correct valuation of the property is an extremely important matter to both the seller and the buyer, and even to the financier. There is no special legislation on valuation or valuers in Finland. Most of the valuations of residential properties are made by the real estate agencies but in addition, distinct valuations are quite frequently performed during the transaction process. To obtain a reliable report, it is important to use professional valuers, e.g. authorised real estate valuers (AKA) (Kasso, 2001, pp. 558-570).

Generally, valuers' fees do not depend on the value of the property. The normal fee for residential property varies between €500 and €1,000 but may be higher if

---

agreements with most countries. The general rule concerning incomes from property is that these are taxed in the country where the property lies (Kasso, 2001, pp. 372-460).

[12] New legislation concerning brokerage has existed since 2001.

the task is more difficult. The fee is subject to VAT, so the value added tax of 22 per cent (year 2002) must be paid in addition to the fee (Viitanen, et al., 2002).

Purchase process

Normally, the process of a purchase of a residential real property begins when the property owner contacts a real estate broker with whom he/she makes a commission for selling the property. With the commission, the owner undertakes to give all information needed by the buyer to the broker. The broker draws an information document for the marketing. Nowadays, it is additionally common for the owner (or broker) to order an inspection of the house (homebuyer's survey) from an authorised inspector before the deal is closed.

The buyer normally gets information on properties for sale from announcements in local journals, on the Internet, or in the brokers' offices. Detailed information of a certain property can be obtained by attending a presentation of the property. The presentations are most often announced in local journals (on Sundays) and they are public. However, private presentations are also arranged. The information document normally includes:

- owner information
- cadastral information (including easements)
- information on local plans
- building information
- mortgage information.

Original documents are presented only when the deal appears to be obvious. A buyer who is interested in the property normally makes a written offer to the seller or to the broker. In addition, he/she normally gives earnest money as a proof of genuine interest. However, according to the Land Code, the offer for a purchase of a real property is not binding unless made in the same procedure as the real purchase contract, e.g. with a public purchase witness. In practice, such an unofficial procedure for offer is regularly used.[13] Should the buyer finally not sign the sales contract after the seller has accepted the offer, the buyer will recover the earnest money but is obliged to pay the cost caused to the seller, or vice versa.

When the seller accepts the offer, the sales contract will be written, normally by the broker. There are, however, no rules regarding who shall or may write the document. The sales contract is signed by the parties concerned and by an official purchase witness. Most often, the contract is signed at the bank that is the financier of the buyer.

---

[13] This is the normal procedure for transactions involving apartments.

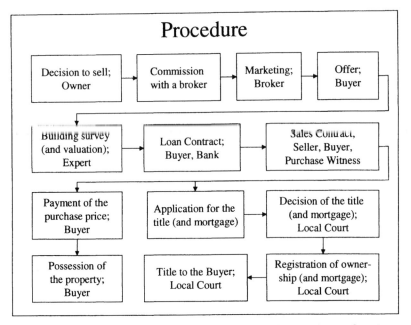

Figure 5:   Scheme of the 'normal' procedure for the purchase of real estate in Finland

The purchase price is normally paid to the seller at the same time as the seller hands over the original documents, including mortgage documents, and keys to the buyer; in fact, the mortgage documents are given to the buyer's bank. As the property to be sold is normally mortgaged for loans of the seller, the seller repays the loans to his/her bank at the same time or provides some other collateral security to the lender. Consequently, the mortgage documents are moved from the seller's bank to the buyer's bank and the money in the contrary direction.

Further, the buyer's bank normally takes care of the registration of the deed and applies for additional mortgages, if required. The commission fee is also paid to the broker, normally by the seller, the cost of the public purchase witness is shared equally between the seller and the buyer, and the registration fees and the transfer tax are paid by the buyer. If the bank or a law firm registers the title, a fee will be paid to them. When there is no need for mortgages, the buyer can easily apply for registration without professional assistance. In any event, it is important to register within six months of the assignment, unless the transfer tax is raised.

The process is also prescribed in Figure 5 and the money process in Figure 6.

Before the sale   **After the sale**

Seller owns a house and
has a loan of €80,000 from
Bank S

Bank S has a
mortgage letter

Buyer has
€50,000

€150,000

Seller gets the sales price but pays the
broker's fee + VAT, survey expert's fee +
VAT, half of the fee of the purchase witness,
and the loan with a service fee €(150,000 –
7,320 – 732 – 50– 80,500 = 61,398)

Broker gets €6,000 (- his/her costs)

Building survey expert gets €600

Public purchase witness gets €100

Bank S gets €(80,000 + 500)

Buyer owns the house but has used his/her
savings and taken a loan of €107,250 from
Bank B to pay the sales price, half of the fee
of the purchase witness, transfer tax, the fee
for the title and fees to the Bank B

Bank B gives a loan but gets €1,000 and a mortgage letter

State gets transfer tax, VAT and title fee €(6,000 + 1,320
+ 132 + 100 = 7,552)

Figure 6: **An example of the money process in the purchase of a built-up one-family house on one's own site (situation in Finland in 2001)**

Conclusions

As can be seen, the drawing up of a sales contract in Finland is not complicated. However, marketing and mortgages assistance from a broker or a bank is normally required. In a normal transaction involving a subdivided plot with a built-up one-family house, the costs for the seller and the buyer are about 10 per cent of the average price. This is met approximately equally by the seller and the buyer.

References

Kasso, M. (2001), 'Asunto- ja kiinteistökauppa', *Transaction of Dwellings and Real Properties (in Finnish)*, 3. p., Kauppakaari, Helsinki.

NLS (2001), 'Kiinteistöjen kauppahintatilasto' 2000, *Real Property Purchase Statistics in 2000*, Official Statistics of Finland, Prices 2001:1, National Land Survey of Finland, Helsinki.

Viitanen, K., Kasso, M., Palmu, J. (2002): *Real Estate in Finland*. Soon coming Publication from Institute of Real Estate Studies, Helsinki University of Technology, Espoo.

Viitanen, K., Vuorio, K., Yli-Laurila, M., Anttila, K. (1997), *Urban Property Market and Land Law in Finland*, Publication B80, Institute of Real Estate Studies, Helsinki University of Technology, Espoo, Finland.

Chapter 5

Property Transactions in the UK: A Situation of Institutional Stability or Technical Change?

Robert Dixon-Gough and Mark Deakin

Introduction

The institutional arrangements concerning property transactions within the United Kingdom are significantly different to those found in the remainder of Europe and for most parts of the world. Even within the United Kingdom, the evolution of the legislative institutions and procedures adopted for property transactions are fundamentally different within the three main jurisdictions of England and Wales, Scotland, and Northern Ireland. An understanding and appreciation of those differences are of fundamental importance in attempting to develop of model of property transactions in Europe.

One of the over-riding factors concerning property transactions in the United Kingdom is the institutional stability of the generic procedures. During the nineteenth century, there was a move to bring those institutional procedures more in-line with those in place throughout most of Europe and, indeed, those adopted throughout the, then, British Empire. The Members of Parliament, who were in the most part both interested parties and members of the legal profession, overturned this proposal in the House of Commons. Conversely, however, those institutional structures that still remain are sufficiently flexible to permit any member of the public to conduct the legal procedures relating to the property transactions, without any legal training.

The registration of real estate properties in the United Kingdom is through the respective land registers. In the case of England and Wales, this is HM Land Register; in Scotland the Land Registers of Scotland, and in Northern Ireland the Land Registers of Northern Ireland. The evolution of land registration in the respective jurisdictions and their respective institutional procedures are related to the Ordnance Survey (Great Britain) and the Ordnance Survey of Northern Ireland. All land registers have a commitment towards the digitisation of their records but, to date, not all properties are registered.

The institutional procedures are currently being reviewed in the light of current technological changes. The main impetus for these changes was the Citizen's Charter of 1992, which gave rise to the introduction of a 'National Land

Information System' (NLIS) for England and Wales, together with corresponding systems for Scotland (ScotLIS), and Northern Ireland (LandWeb). These technological changes in the context of parcel-based property transactions, cannot be discussed without mention to the National Land Use Database (NLUD) and the National Land and Property Gazetteer (NLPG).

The historical evolution of the institutional procedures

The institutional processes of land and property transactions in Great Britain have gradually evolved over a period of some 800 years. The same cannot be said for Northern Ireland since many of those evolutionary processes were not available to the Irish people.

Powelson (1988) suggested that in many of the nations of northeastern Europe, particularly England and Wales, there has been the opportunity to develop feudal arrangements over time through the evolutionary process of contract feudalism. Negotiations have taken place over centuries between landlords, tenants, and commercial factions that have enabled the participants to gain equality in status through the development of a democratic form of government.

England and Wales

The evolutionary process of land reform commenced in England in 1215 with the signing, by King John, of the Magna Carta. In forcing the King to sign this document the Barons were not strong enough to oppose him but had to seek the aid of the church and all classes who had been oppressed by the Crown. The terms of the Magna Carta were of limited interest to 'the people of the nation' in the thirteenth century, but:

> owing to the economic and legal evolution of the next three hundred years it came to embrace the descendant of every villein in the land, when all Englishmen became in the eye of the law 'freemen' (Trevelyan, 1988).

This process continued in England and Wales with the Black Death of the fourteenth century (and the ensuing shortage of labour) and the definition of the Rights and Prerogatives of the common man in the 15th century, which gave rise to the voice and needs of the rising middle classes. The War of the Roses eventually led to the breakdown of feudalism, which was essentially due to the decline in the number of villein tenants. This, in turn, resulted in competition for workers amongst manorial lords, bringing into being the class of agricultural labourer prepared to hire out their service to the highest bidders. A further evolution came during the middle of the fifteenth century, when payments were made in the form of ground rents for their holdings rather than in the feudal burden of personal services.

As the social status of villeins improved, their tenure became known as Copyhold. This expression was the result of their title to land being recorded in

Court Rolls, the villein tenant being said to hold a 'copy of the court roll', the copy being his title deed. Since all transactions were recorded upon the roll, it provided conclusive evidence of the copyholders' rights (Simpson, 1984). In the nineteenth century, Torrens suggested that the British missed the opportunity of setting up a National Land Register based upon the Copyhold Laws. By that time, several statutes had been passed to encourage the voluntary extinguishment of the copyhold tenure. Copyhold was compulsorily abolished in 1925, tenures being transferred as a result of the Land and Property Act.

During the period of the industrial revolution, one of the greatest periods of social change took place throughout Great Britain. Prior to the industrial revolution, the principal status of power was land and, in order to keep land, many of the families of the large estates tied up the land through inheritance to prevent it from being sold by future generations. The key value of the estates depended upon them being kept intact. Thus, their total value represented a status that was greater than the sum of their parts. Ownership represented influence, which governed many aspects of local life. The industrial revolution changed this and shifted power to the factory owners who required land both for factories, their workers houses, and for their own status.

The sale of land was complicated because of the English Land Laws. Furthermore the process was time-consuming and resulted in an expensive process of private conveyance and risk of fraud. These complications lead to the decline in land values at a time when the market for land was increasing. Concern at the increasing confusion and insecurity of title led, in 1828, to the appointment by Parliament of a Royal Commission to inquire into the law of England and Wales concerning Real Property.

This period also witnessed several significant steps in the registration of land in England and Wales. The 1830 Royal Commission resulted in four reports, the first two devoted to land registration (1829 and 1830), the third to land tenure (1831-1832), and the fourth to wills (1832). In their main report, the Commissioners concentrated upon the gross defects of the prevailing methods of land transfer, and on the need to establish 'a General Registry of Deeds and Instruments Relating to Land'. They recommended that a Deeds Registry should be established with a general office in London serving England and Wales, together with regional District Registries. The Commission took it for granted that the registration of land would be compulsory and this was accepted at the time by Parliament. Had this been achieved, the register of land in the United Kingdom might now have been complete (Riddall, 1983). This resulted in the introduction of a General Register Bill (1830) in the House of Commons, which was opposed largely by solicitors who feared a reduction in their business. Despite the failure of the bill, it was a pioneering document in that it made reference to the Registration of Title as distinct from the Registration of Deeds.

Apathy on the part of the public and opposition on the part of the legal profession resulted in a lack progress being made on the registration of land. This led to the report of the Registration and Conveyancing Commission, published in 1850. In 1846, a select committee of the House of Lords came to the conclusion that the marketable value of real property was being seriously diminished by the problems associated with its transfer. A Royal Commission was appointed in 1847 to explore the most effective system for the registration of deeds to simplify and reduce the costs of conveyancing. The Commission concluded that the existing rights to land were too complex for the title to be guaranteed to a public officer but recommended, however, the setting up of a National Deeds Registry. The recommendations of this report formed the basis for the current land registration in England and Wales. The previously requested deeds register was rejected in favour of the more radical concept of a General Registration of Title to Land.

In 1853, the Ordnance Survey commenced the 25 inches to the mile survey (approximately 1:2,500) of rural areas and 50 inches to the mile of urban areas (approximately 1:1,250) in England, Wales and Scotland. This was the first systematic mapping of 'visible' boundaries in Great Britain and was not completed until 1893. It did, and successive series still do, form the basic tool for land registration in Great Britain. The Ordnance Survey of Ireland produced a similar series of maps.

In 1862, the Land Registry of England and Wales came into being. Registration was voluntary and the records kept secret. Only the owners of estates and interests were allowed to inspect the register and then only their own property. Further recommendations led to the 1875 Land Transfer Act, which was voluntary and once more, not a success. In an attempt to make land registration more attractive to land owners, three new elements were introduced; land had to be sold before being registered, boundaries no longer had to be accurately surveyed but recorded as general boundaries and, thirdly, partial and equitable interests could no longer be registered. The definition of a general boundary means that the exact line of the boundary is left undetermined. Instead the system makes clear where the parcel is situated in relation to certain clear and visible features, such as hedges. This was a re-introduction of a rule that had been applied to conveyancing for centuries. The Land Registry was not empowered to make its own maps, the Land Transfer Act simply defining that the applicant provided an extract from a local map, which had been defined as the 'public map of the district'. It was, therefore, reliant upon Ordnance Survey maps. In the absence of these, tithe maps and estate plans were used.

The Compulsory Land Registration Act of 1897 was introduced for selected areas, initially in the County of London and extended to the City of London in 1902. In order to pass the Act through Parliament, two important amendments were necessary. Essentially, the Act called for the compulsory registration of Title on the sale of a property, or upon granting a lease for forty years or more. The first amendment allowed the Act to be introduced as an experiment within the Administrative County of London. The second amendment came with an Order making registration compulsory only in areas where the County Council requested it. The Register was to remain secret. The 1897 Land Transfer Act defined the Ordnance Survey plans as the basis of all descriptions of registered land. The map became dominant so that if any verbal descriptions of the property disagreed with the plan, it was the graphical description that was to take legal precedence.

During the latter part of the nineteenth century a number of Finance Acts were passed that taxed the value of inherited land holdings. These evolved into the 1910 Finance Act for the redistribution of wealth from primarily landed gentry. This was referred to as The People's Budget introducing, amongst other things, 'death duties' later to be superseded by the Inheritance Tax. These Acts, which included the introduction of income tax, led to a general decline in the wealth, status and of the influence of the landed gentry. However, the transfer of property from the ownership of those estates did not reach any significant level until the demand and value of land was revived during the period between 1911 and 1914. During and immediately after the First World War, many large estates had to be sold and broken up to pay for death duties brought about by the decease of two or three generations in rapid succession. It has been estimated (Thompson, 1963) that from 1918 to 1921, between six and eight million acres changed hands in England and Wales, some 25per cent of the total land area. This lead to a breakdown in systematic agrarian planning and rural land management for virtually the remainder of the 20th century (Mansberger *et al*, 2000). Thompson (1963) commented that these land transfers:

> marked a social revolution in the countryside, nothing less than the dissolution of a large part of the great estate system and the formation of a new breed of yeomen.

The 1925 Land Law Legislation evolved from recommendations made by the Acquisition and Valuation of Land Committee, which met in 1919. The principal recommendation of the Committee was that the provisions of the Act of 1897, under which compulsory registration was impossible except at the instance of a County Council, be repealed on the grounds that the gradual extension of compulsory registration was of a national rather local interest. The 1925 Law and Property Act lead to the gradual introduction of compulsory land registration across England and Wales. Despite subsequent legislation in 1960, a considerable number of land parcels remain to be registered.

Scotland

Land in Scotland comprises the air, sea, and inland water within the territorial limits. In addition to its physical limits, this territory is also a political concept since it defines the extent of sovereignty and jurisdiction of Scotland's land law. This land law establishes the legal framework for land ownership, the legal basis being the land tenure system, which remains feudal. Feudalism is characterised by a hierarchical system of rights, with God and the Crown at the apex and superiors and vassals below them. The rights of feudal superiors including the Crown therefore qualify title to land under feudalism. Traditionally, and before the Act of Union (1603), the laws of Scotland had been different to those of England and Wales. The foundation of modern Scots law was provided by the publication of the *Institutions of Laws of Scotland* in 1681, and since then there have been strong influences from the continent of Europe and from England.

Land law in Scotland is still, in theory, feudal (Doughty, 1999). Feudalism is commonly accepted as being introduced in the 11th century, during the reign of King David I, and was originally conceived as a system of government in which the Crown granted rights to the nobility in return for military and financial services. This has gradually been converted into a system of private property with the Crown now having limited powers. All landowners are vassals of the crown, but some are more superior to others through the retention of certain rights, or the imposition of certain obligations on those to whom they sell their lands. These purchasers or vassals can, in turn, become superior to others through the same process (Wightman, 1996). The public interest has been increasingly expressed through statute law, which for example, has nationalised planning permission and certain mineral rights. Although many of the traditional roles of superiors and vassals have disappeared, the old practice of feudal service was transformed into a perpetual payment known as feu-duty.

Whilst the most prominent form of land tenure in Scotland is feudal tenure, land may also be held alloidally – independently with no feudal superiors. One specific type of alloidal tenure is Udal tenure, which exists in the Orkney and Shetland Islands and is a product of the islands' Viking heritage. Udal tenure provides for the inheritance of the property by all of the owner's children rather than simply the oldest son or child.

One of the reasons why feudalism has persisted in Scotland is that it has provided a seamless property system for over 1,000 years and has affected with it a sense of status, real control, and political power (Wightman, 1996). Feudalism was abolished on the continent largely as the result of revolution and Napoleon. However, in Scotland the influences of generations of feudal Barons have made its Civil Law traditions seem out of step with the traditions of mainland Europe (Burdon, 1998).

Scotland was the first country in modern times to establish a public land registry for the protection of land rights (Pitticas, 1992), which unlike those of England and Wales have been open to public inspection since their inception. The register, which dates from 1617, was designed to end violent disputes over property. It is referred to as the *Register of Sasines*, derived from the French word *seisin,* which means to seize or to hold. This concept is in sharp contrast to the Land Register of England and Wales, which until 1990 was only open to owners and agents (Burdon, 1998).

Since the 17th century, the registry has remained largely unchanged and every piece of privately owned heritable land in Scotland is recorded in the Register held in Edinburgh. Whenever a property is created or sold, it is updated. It therefore holds the ownership, history, sales, and mortgage records of over a million properties and is amended some 400,000 times each year (Lloyd, 1996). The value of the Register as an information source is somewhat limited by the indexing system. It is not map-based and thus the knowledge of either the person involved or the name of the place concerned must be the basis for any search. This leads to a search sheet, which contains the pertinent information about the subject panel. These are only descriptive but may be cross-referenced to plans. Difficulties may arise when there is multiple ownership or when the parcel is unusually large, making the acquisition of all available information a very tedious process. The system is, however, more than simply a register of deeds. In 1844, the Court of Session ruled that title was not complete until the fulfilment of recordation. Thus, recordation became an integral and essential component of the transaction process, without which the grantee cannot assume full title (Burdon, 1998).

The process of maintaining the Register of Sasines is expensive due to the length of the deeds and the amount of time needed to access and examine the register. The Registration of Title was introduced as the result of the Land Registration (Scotland) Act 1979, in an attempt to make the process more efficient. It was first applied in 1981 to the county of Renfrew and differed from the Register of Sasines (which records the evidence of title) by guaranteeing that good title will be delivered to the purchaser (Pitticas, 1992). In order to meet this goal, there had to be a very thorough legal examination of the history of each property at the time of registration. Initially, it was planned to introduce the new registry to the whole of Scotland by 1990, but the coverage was limited to about 30 per cent of the country due to resource and budget problems. To improve this, the Registers of Scotland (RoS) sought to introduce new technology and improved procedures in an attempt to develop increased flexibility and efficiency. The functionality of GIS was utilised through the incorporation of the land registry into the agency's digital mapping system and by 1998, some 14 of the 33 counties had become functional, with the remaining expected to become functional by 2003. Those 14 counties represented the majority of the populated areas, with the remaining (mainly rural) still under the Register of Sasines (Doughty, 1999; Burdon, 1998).

The Keeper of the Registers of Scotland delivers to the grantee a title sheet known as a land certificate, which consists of four sections: property, proprietorship, charges, and burdens. The property section contains a brief description of the property, its postal address and a reference to the relevant OS map. The proprietorship section contains the name of the owner, the proprietor's heirs, the price paid, the date of registration, and the date of moving into the property. The charges section contains details of the mortgage on the property, whilst the burdens section list any restrictions on the property. Whilst the new Land Register improves the efficiency of conveyancing and reduces the likelihood of title defects, there still persists the common complaint that it is very difficult to simply and inexpensively determine who owns what parcel of land.

Northern Ireland

The system of land registration used in Northern Ireland is loosely based upon the Torren system, but more closely related to the 'English System' (Henssen, 1995). The system consists of a registry of deeds, a statutory charges register, and a land register. These registers, particularly the Land Register, provide a mechanism for the government of Northern Ireland to guarantee title for registered property.

The history of Ireland is very turbulent, with Cromwell's repression of the 1642 O'Neill uprising, leading to the expulsion of the Roman Catholic Irish from Ulster.

Through this process, the indigenous Irish became disenfranchised from their lands, eventually leading to a feudal landlord/serf relationship between the English and their Irish tenants. Over the next 350 years, several Acts of Parliament gradually restored many of the rights to the tenant farmers and land reforms were implemented, particularly during the nineteenth century. Two examples of such Acts included the Landlord and Tenant Law Act (Ireland) of 1860 and the Land Purchase Acts of the 1880s. The former effectively eliminated the feudal relationship between landlord and tenant and transferred the relationship to a contract-based landlord and leaseholder arrangement. The Land Purchase Acts were instrumental in providing the Irish tenants with secure land tenure, whilst giving leasehold tenants the rights to purchase freehold interest in the land or simply to gain an interest in the land. In most instances, however, the estate was converted into various lease devices, such as the 'fee-farm grant', a type of fee simple tenure in which the land is held in perpetuity in exchange for an annual rent (Wylie, 1995). This type of tenure grants secure tenure to the former tenant with no reversion to the former landlord. Other forms of leasehold relationships also exist in Northern Ireland including leases for lives (for a certain number of years or for perpetuity), rights of residence (life estate) and conacre (a form of occupancy where licence is given for the use of the land for a specific purpose, typically farming).

These various arrangements have caused the system of land tenure in Northern Ireland to be very complex, especially in the urban areas. Whereas in rural areas, the Land Purchase Acts have led to the creation of single fee landowners, more complex pyramids of titles are very common in cities and urban areas (Thomas, 1997). As properties are leased and sublet, it has proved to be very difficult to determine the status and identity of those holding title to many of the properties in urban areas. It is for reasons such as these that the Land Registry of Northern Ireland can only guarantee title on approximately 50 per cent of properties. This has the effect of making land transactions more difficult since the title search is performed through the Registry of Deeds.

The cadastral system of Northern Ireland is based on the large-scale topographic plans of the Ordnance Survey (Northern Ireland) and relies upon the recognisable physical demarcation of property boundaries (Barr, 1985). Certification of title in the Land Registry of Northern Ireland has been based upon these maps, whilst the registration of land tenure is recorded in three separate registers; the Register of Deeds, the Land Registry, and the Statutory Charges Register. All registers are open to public access.

The Register of Deeds for Ireland was initially established in 1708 and permitted the recording of deeds, or deed abstracts as evidence of land ownership. The Register is indexed by name and all searches must be made by the owner's name. It is not necessary for complete copies of deeds to be held in the register since the function of the register is primarily to acknowledge their existence, without guaranteeing the validity and legality of the document. However, in the case of adjudication, a fully recorded deed will take precedence over an abstract.

The Land Register was adopted in the Act for Ireland (1891). This Act authorised the creation, in 1892, of a Land Register to certify titles established as the result of the land reform acts of the nineteenth century (TCD, 2000). After a parcel had been accepted into the Land Register, the government guaranteed the title. Once the first-time registration has been made, future transactions are no longer recorded in the Register of Deeds. The Land Registry issues several types of title, including absolute title and possessory title, together with entries such as fee farm, fee simple and tenancy for lives. The Land Register also records interest in land holdings, such as rights of way, easements, and mortgages.

The Statutory Charges Register was established during 1951 to provide purchasers with a means of checking whether a property is affected by statutory restrictions that would not otherwise have been easily discovered. This is also a map-based registry.

Changes to institutional procedures

In the case of the three major forms of jurisdictions relating to property transactions within the UK, there are significant changes currently taking place that will have an affect upon the institutional procedures of Scotland, and England and Wales. There are also minor changes in procedures in Northern Ireland.

Land reform in Scotland

A central feature of the legislative programme for the Scottish Parliament's inaugural session was a comprehensive programme of land reform, in which the abolition of the feudal system in Scotland was a key focus. Feudal tenure was first recommended for abolition over 30 years ago. Three of the eight Executive Bills of the 1999/2000 Scottish Parliamentary Session dealt with land reform. The bills included the abolition of the feudal system of land tenure; the creation of Scottish national parks; and the statutory foundation of a 'right to roam' and of communities 'rights to buy' (Scottish Parliament, 2000).

The principal bill, commonly known as the Abolition of Feudal Tenure (Scotland), was intended to replace the feudal system of land tenure with a system of simple ownership. Feudal superiorities will disappear, with land in the future being owned outright. The abolition of the feudal system of land tenure in Scotland is widely recognised as being a matter of sensible necessity. In practical terms, however, remnants of the feudal system currently survive.

The new Land Registration Act of England and Wales

The Land Registration Bill was passed by the House of Lords on the 8th November 2001 and is expected to become operative as an Act of Parliament during 2003. This is the result of 6 years work by the Law Commission and the Land Registry that resulted in *Land Registration for the Twenty-First Century*, published by the Law Commission (www.lawcom.gov.uk).

The new bill will enable electronic conveyancing, change the rules on adverse possession, create a new office of Adjudicator to HM Land Registry, extend the rules governing first registration, and simplify many of the complex issues of interests, priorities and charges. The new bill is intended to supersede the 1925 Land Registration Rules and will specifically create a framework to permit the transfer and creation of interests in land by electronic means. Both electronic conveyancing and electronic searching will be progressively and gradually introduced through the set-up of secure communication networks and it is anticipated that all conveyancing will eventually be conducted electronically.

The Bill also proposes that the Lord Chancellor appoint the holder of a new post of Adjudicator to the HM Land Registry. The post-holder will hear and resolve disputes that parties bring to HM Land Registry rather than to the courts. Significantly, the triggers for first registration will also be extended as it attempts to stimulate more first registrations, and to allow leases with 7 years or more to run (as opposed to the current 21 years) to become registered. Through the registration of *profits à prendres*, the registration of certain crown land and submerged land, it will also ensure that the register is more complete.

Northern Ireland

In 1995, the Compulsory Registration of Title Order (Northern Ireland) was passed and applied to two areas (County Down and County Armagh) to ensure that all land conveyances resulting after 1st June, 1996 are registered in the Land Registry. This includes property inherited through a will and any property for which a mortgage is sought. Through the introduction of this order, the government hoped to gradually bring into the Land Registration system, all previously unregistered property.

The Property (Northern Ireland) Order of 1997, which came into operation in January 2000, brought about fundamental changes to property law in an attempt to streamline the complicated nature of title to property in Northern Ireland (CFR, 1999). From the 10th January 2000 several forms of deed were abolished including fee farm grants, and long leases.

Technical change

The concept of a National Land Information Service (NLIS) for the UK was initially proposed by Professor Dale at the AutoCarto Conference of 1988, in which he envisaged fast and easy access to a comprehensive record of all land and property (McLaren and Mahoney, 2000). The origin of the NLIS can be traced back to the Citizen's Charter of 1992, the central theme of which was that the creation of:

A national land information system may be one way to allow the citizen faster and easier access to an authoritative and comprehensive public record of all land and property.

Both the NLIS and its Scottish equivalent (ScotLIS) were initiatives intended to improve access to land and property related information. The overall concept of the two initiatives is to provide an electronic delivery of land and property-related information, on a parcel basis, to a wide audience (Deakin, 1998, 1999; Ralphs and Wyatt, 1999). The initial target, however, is to provide search facilities for land and property information to support the process of conveyancing. These two initiatives have played an important role in promoting the significance of spatial data to both government and the business community (Smith and Puddicombe, 1998), whilst recent political changes have proved that spatial data can be used to support policy decisions (Mahoney and McLaren, 1999).

At the centre of NLIS/ScotLIS is the National Land and Property Gazetteer (NLPG), which has standardised the way by which all geographical addresses are designated. This will be through a unique property reference number (UPRN) to which all data related to that property, in any database, can be linked. Given that every property will have a unique descriptor, it will be possible to adopt a single property search that will be linked to all relevant databases. Thus, the NLIS/ScotLIS effectively involves the linking of data sets to geo-spatial information with the intention that the process of property searches can become a 100 per cent paperless procedure. Each property will have an agreed address, a grid co-ordinate, possibly spatial dimensions, and will be assigned a unique property reference number (NLIS, 2000). It will have the long-term potential for linking all types of geo-spatial data sets, such as retail information and crime data.

Prototype and pilot projects

In the case of England and Wales, the trial area consisted of two postcode areas in the city of Bristol. At the centre of the project was the development of the central land and property gazetteer to control all property-related information. The Council was, at that time, also involved in the Local Government Management Board's development of the BS7666 (Spatial Datasets for Geographic Referencing) and the Gazetteer was developed to meet the needs of both projects. With information supplied by HM Land Registry (HMLR), the Ordnance Survey, the Valuation Office, and Bristol City Council, a Land and Property Gazetteer (LPG) was set up in 1995 to support property transactions. Through the use of a demonstrator, the Local Land and Property Gazetteer (LLPG) was used to provide on-line access to data from the four organisations.

The pilot NLPG was used to service search requests for the solicitors and three practices in the Bristol area were provided with on-line access to the NLIS services with searches being submitted and returned electronically. They had access to both digital mapping and LPG data to enable accurate property identification. According to Musgrove and Yeoman (1998) the:

Gazetteer forms part of the foundation of the process as it provides the essential link that spatially enables land and property-related information, allowing it to be accessed or manipulated in conjunction with geographical or map-based data.

The project was extended to cover the whole of the city and used to provide users with map-based, definitive property identification to users of the system.

The development of a pilot ScotLIS demonstrator began in September 1995, its main objective being to raise awareness and interest in the concept of the project through the development of a demonstrator system integrating land and property information data sets from a wide range of data providers. A secondary objective was to investigate issues relevant to the creation of on-line access to data sets and to test the address matching algorithms and gazetteer creation tools developed by the OS and utilised within the NLIS Bristol trials.

Following the success of the pilot projects conducted in England and Scotland, it was announced in 1999 that a new electronic land registration system (LandWeb), designed to improve the efficiency of conveyancing, would be implemented in Northern Ireland (NIIS, 1999). This will allow solicitors and money lending institutions access to information and to submit applications from their own offices through secure links. The system will also make it easier for members of the public to obtain information about land through map searches (NLRI, 1999).

Improvements to be introduced by the NLIS

At present, for the conveyancing of every property within the UK a search has to be made relating to the property being bought. The acting solicitor (or individual) directs this to the appropriate local authority, and sets in motion a labour intensive process that can take several days or weeks to complete. Although many local authorities have their own integrated information systems that can short-cut many of the processes, much of the search will still involve a land charge officer who has to ask specific questions to all relevant departments, both at local and national level.

Much of this will be avoided when the NLIS comes fully on-line since it will permit access to a wide range of records, such as the NLUD, providing such information as past land use, and risk of contamination, flooding and subsidence.

In December 2001, a statement was made from the Lord Chancellor's Department to the effect that approval had been given by the Government for the establishment of an interdepartmental working group, known as the Home Buying and Selling Task Group. One of the key issues to be addressed was that of *E-conveyancing*, which is considered to be the catalyst for the re-engineering of national property transaction systems, since it will provide better conveyancing services to both businesses and to individuals. HMLR, which already provides many services electronically, will make all of its key services available online from 2005. In addition, the Land Registration Bill will provide a legal framework for a new electronic conveyancing system, which is targeted for operation in 2006.

Cadastre by stealth?

Her Majesty's Land Registry (HMLR) currently operates a register of titles to land and interest to land. This consists of over 18.5 million registered titles, each indicated on its filed plan, based on the largest scale OS mapping. The Index Map is the index to all plans that have been filed and is presently held on paper as a series of some 400,000 extracts of OS maps, which are located in the 24 District Land Registers and updated manually.

One of the most important features of land registration in Great Britain is that it is based on the general boundaries rule, by which the extent of the land parcel is defined by reference to the physical features that surround it. This means that the precise positions of the legal boundaries of the land are not defined. Thus a physical boundary might be a wall, but whether the boundary follows the centre line, or either side of the wall, is not defined.

HMLR has proposed that its index map be digitised for a number of strategic reasons. Firstly, it will provide their staff with on-line access to the Index Map to provide an electronic gateway to their definitive title information. Secondly, it will allow more efficient processing of registrations and enquiries and, thirdly, it will eventually be made available for public viewing and will contribute towards electronic conveyancing applications within the NLIS. This action has led to the suggestion that this action might be likened to the introduction, by stealth, of a numerical cadastre for Great Britain. However, according to Dale (1976), numerical cadastres normally have the following features: fixed boundaries, monuments at every turning point along the boundary, precise surveys of every boundary monument, and plans showing bearings and distances between boundary monuments. The HMLR digital Index Map will possess only one of these features – the boundary co-ordinates, although it would be easy to derive bearings and distances from these. Because of these limitations, it would be difficult for the digital Index Map to act as a numerical cadastre. Maynard (2001) has identified four important factors that would prevent the digital Index Map from being used in this form. These are; the quality of the information given to HMLR on which the first registrations were based (often copies of the relevant OS plan with the boundaries crudely identified with a red pencil), the rules by which HMLR interpret the information they were given, the accuracy of the underpinning Ordnance Survey map, and the lack of fixed boundaries in Great Britain.

As a result of these factors, Maynard (2001) argues that the digitisation of the Index Map cannot be viewed as step towards a numerical cadastre since it is nothing more than a digital image of a record of the 'general boundaries' registered titles.

Role of the surveyor

The role of the surveyor in the institutional procedures of property transactions throughout the United Kingdom remains tenuous and, in common with the role of the legal profession, not always necessary. The surveyor is responsible for two main roles, both of which may be obligatory from the part of the institution financing the purchase. The first role is that of assessing the security, stability, and integrity of the land or building, i.e., a surveyor concerned primarily with building structures. Although this role may be viewed by the purchaser as an assessment of the soundness of the building and land, the building surveyor primarily provides a report to satisfy those providing the finance that the property will provide adequate collateral for the loan. Therefore, part of this role may also involve some degree of valuation. The second role of the surveyor is related purely to the valuation of the property and land, which might overlap or be performed in co-operation with the building surveyor. It is essentially an assessment of market forces. Once again, this process is not obligatory but may be insisted upon by those financing the purchase of the property. The function of valuation can be in connection with securing a loan with which to purchase the property, but might also be in connection with taxation, in the case of securing probate to an inheritance, or even in the compulsory purchase of the property.

Unlike most countries throughout the world, the role of the licensed or cadastral surveyor does not exist within the United Kingdom. This is for two main reasons. Firstly, the concept of a cadastre is not accepted and, secondly and partly as a result of this, the system of general boundaries is used. The main role of the land or topographic surveyor in property transactions throughout the United Kingdom, has been the provision of large-scale topographic maps depicting the position of those boundaries that are in evidence at the time when the property was surveyed. The position of those boundaries is commensurate with the plotting precision and accuracy of the scale of representation. In urban areas, this has been 1:1,250 (or imperial equivalent), 1:2,500 (or imperial equivalent) for rural areas, and 1:10,000 (or imperial equivalent) for areas of mountain and moors. The map detail is used to define the general property boundaries that are also described in detail in the title deeds. The outline of the boundary, as it appears on the largest-scale OS map, is also used to define and register the property. None of the actions required in defining the boundary relies upon the action of a surveyor.

Conclusions

The institutional procedures that govern property transactions throughout the UK have evolved over long periods of time. There are significant differences within the three jurisdictions of the UK, but there is a strong generic root running through all procedures. Although those procedures are relatively complicated, their institutional stability is such that they are sufficiently flexible to permit transactions to be conducted both by representatives of the legal and surveying professions, and by private individuals.

As a result of technical change and the increasing availability of other relevant forms of geo-spatially referenced data, the three jurisdictions have addressed the need to provide systems relating to property transactions, e.g., the Land Registry, to be opened up through the National Land Information Service and in so doing, bring the institutional procedures more in-line with those practised across Europe.

References

Barr, M., 1985. *Comparisons among Land Records of the European Colonies and Other, Improved Systems Used in Developing Countries,* World Bank, Washington D.C., USA.

Burdon, I., 1998. *Automated Registration of Title to Land,* Registers of the Scotland Executive Agency, Edinburgh.

CFL, 1999. *Major Changes in Northern Ireland Property Law - 10th January 2000,* Cleaver Fulton Ranking.

Dale, P.F., 1976. *Cadastral Surveys Within the Commonwealth,* HMSO, London.

Deakin, M., 1998. The development of computer-based information systems for local authority property management, *Property Management,* 16, (2), 61-82.

Deakin, M., 1999. The development of computer-based information systems for local authority property management. In: Deakin. M., (ed.), *Local Authority Property Management: Initiatives, Strategies, Re-organisation and Reform,* Ashgate Publishing, Aldershot.

Doughty, S.W., 1999. *Land tenure and crafting in Scotland,* Department of Spatial Information Science and Engineering, University of Maine.

GM, 2002. New Land Registration Act for 2003, *Geomatics World,* 10(2), 6-7.

Larner, A., 2001. A fully connected land is in sight, *Public Service Review,* Autumn 2001, 38-39, Public Service Communication Agency, London.

LCD, 2001. *Statement by the Lord Chancellor on the work plan for taking forward HM Land Registry's quinquennial review recommendations,* Lord Chancellor's Department, 12th December 2001.

Lloyd, C., 1996. Keepers of the house: a new land register for Scotland, *Mapping Awareness,* 10(5), 40-43.

LRNI, 1999. *The Land Registry Map,* Land Registers of Northern Ireland, Department of the Environment for Northern Ireland, Belfast.

Mahoney, R.P. and McLaren, R.A., 1999. The use of spatial data to enhance the democratic process. Paper presented at the *FIG Commission 3 Annual Meeting and Seminar,* Budapest, Hungary.

Mansberger, R., Dixon-Gough, R.W. and Seher, W., 2000. A comparative evaluation of land registration and agrarian reform in Austria and Great Britain. In: Dixon-Gough, R.W. and Mansberger, R., (eds.), *Transactions in International Land Management,* 1, 73-104. Agate Publishing, Aldershot.

Maynard, J., 2001. Digital boundaries in England and Wales – a numerical cadastre? *Surveying World,* 9(4), 1-7.

McLaren, R.A. and Mahoney, R.P., 2000a. Breakthrough in revolutionary services for land and property transactions. Lessons learned in implementing a NLIS in the UK. Paper presented at the Athens 2000 Workshop of FIG Commission3, *Spatial Information*

Management Experiences and Visions for the 21st Century, Technical University of Athens.

McLaren, R.A. and Mahoney, R.P., 2000b. NSDI in the UK, *Quo Vadis - Proceedings of the International Conference*, FIG Working Week, Prague.

Musgrove, T. and Yeoman, B., 1998. NLIS – the vision becomes reality. Paper presented at the *Association of Geographic Information* Conference.

NIIS, 1999. *Minister Announces Online Conveyancing Breakthrough*, Northern Ireland Information ServiceNLIS, 2000, *NLIS in Depth*.

Pitticas, N., 1992. Scotland the brave = leaping from land registration to land information, *Surveying World*, 1(1), 31-33.

Powelson, E., 1988. *The Storey of Land: A World History of Land Tenure and Agrarian Reform*, Lincoln Institute of Land Policy, Cambridge, MA, USA.

Ralphs, M. and Wyatt, P., 1999. The application of geographic and land information systems to the management of local authority property. In: Deakin, M., (ed.), *Local Authority Property Management: Initiatives, Strategies, Re-organisation and Reform*, Ashgate Publishing, Aldershot.

Riddall, J.G., 1983. *Introduction to Land Law (3rd Edition)*, Butterworths, London, England.

Scottish Parliament, 2000. *Abolition of Feudal Tenure Etc. (Scotland) Bill*, Research Paper 00/09, The Scottish Parliament Information Centre, Edinburgh.

Simpson, R.S., 1984. *Land Law and Registration (2nd Edition)*, Surveyors Publications, London.

Smith, R.J. and Puddicombe, A.G., 1998. A national land information service - moving from concept to reality. Paper presented at the *21st FIG International Congress*, Brighton.

TCD, 2000. *The Universal Land Registration Project, Gazette of Land Registration Systems,*

Thomas, J.H., 1997. A History of Land Tenure Arrangements in Northern Ireland, Department of Spatial Information Science and Engineering, University of Maine.

Thompson, F.M.L., 1963. *English Landed Society in the Nineteenth Century,* Routledge and Kegan Paul, London.

Trevelyan, 1988. *A Shortened History of England*, Penguin, London.

Wightman, A., 1996. *Who Owns Scotland*, Cannongate Books, Edinburgh.

Wylie, J.C.W., 1995. The Irishness of Irish land law, *The Northern Ireland Legal Quarterly*, Autumn/Winter.

Chapter 6

Land Tenure and Real Property Transaction Types in Latvia

Armands Auzins

Abstract

This article has a descriptive character with some analysis of land tenure and transaction types in Latvia. The impact of societal background on land tenure and its relation to the types of transactions is reflected as well. The forms of land tenure mentioned in the article become valid during the transition period.

The purpose of this contribution is to identify the ongoing activities and initiate proposals for development of a platform in Latvia would help to reach the main objective of the COST G9 action – to improve the transparency of the real property market and to provide a stronger basis for the reduction of costs of real property transactions by preparing a set of models of real property transactions which are appropriate, formalised and complete according to stated criteria, and then to assess the economic efficiency of these transactions (COST 328/00, 2001).

Such a platform should serve here as infrastructure of real property rights by providing the required basis for the relevant legislation, stabilisation of societal background, informing and involving of society, improvement of skills and relationships of both professionals and authorities, provision of the required material basis, implementation of a future development concept, and other relevant activities.

Introduction

After the renewal of independence of the Republic of Latvia and the disintegration of the planned economy system in 1990, a transfer was made to a market economy. Through the implementation of the land reform, legal, social and economic transformations started in urban and rural areas.

The aim of the land reform in Latvia has been to implement a step-by-step process of denationalisation, conversion, privatisation of illegally expropriated properties, to reorganise legal, social and economic relationships concerning land property and land use in order to facilitate the development of infrastructure, land protection and rational land use in the interests of society.

It is difficult to say whether the process of the land reform has proceeded according to the above mentioned goals, but it is worth mentioning that the scope of problems is rather wide and a large amount of work has been done up till now. The main reason why the land reform cannot be completed is the lack of financial resources – both by the State and by members of society. It is expected that approximately 20 per cent of individuals and legal persons, who enjoy the right to use land, will not gain land ownership rights during the land reform.

In practice, the land reform is carried out in three directions: restitution of real property rights, privatisation of real estate and compensation for previous ownership.

Real estate can be privatised and compensation for previous ownership paid by using privatisation vouchers. The nominal value of one voucher is 28 LVL (lats, the Latvian currency); the amount equal to approximately 46 USD, but its market price is much lower.

Privatisation of land is going on, and a land market has begun functioning. 42.8 per cent of the total land area and 60.1 per cent of non-agricultural land were in private ownership as of January 2001. In order to accelerate the process of land privatisation, the required funds have been allocated, and the regulations governing the land privatisation process have been simplified. The State Cadastre of Real Estate is functioning, and, with 589,000 land properties recorded, 98 per cent of all land registration cases have been completed.[14] The National Computerised Land Book comprising a central register and a data transfer system became operational in July 2001.[15] By June 2001, 527,137 property units were registered in the Land Book. However, there is widespread dissatisfaction as regards the lengthy process of registration in the Land Book (CEC Regular report, 2001).

When examining both processes, i.e. privatisation and transactions, there seems to be no interrelation, but the process of the land reform influences the creation and functioning of the real property market very profoundly.

Through the process of privatisation, the rights in real property and obligations associated with such property (land with buildings or without, buildings or flats) are transferred from the State to the either individuals or legal persons in compliance with specially drafted legislation – regulations predominantly concerning the land reform. Therefore it is possible that different persons own a land parcel, a building on this land parcel, and a flat within this building. In other words, there is divided ownership.

Each real property that has been privatised is considered to be an object of future transactions. The smoothness of this future process depends on how technically and legally correct and economically substantiated privatisation of real properties has been and how the property rights have been restituted, and what were the consequences. We should not forget that a change from one form of land

---

[14] The State Cadastre of Real Estate contains information about ownership, land use, buildings, and value.

[15] The Land Book defines ownership of a property, it records encumbrances and changes to property ownership, and it contains information about ownership, mortgages, easements, and encumbrances.

tenure to another goes through certain procedures that take time and involve costs. For instance, a change must be made from free state land or land use rights to ownership rights.

The development of a sustainable platform or infrastructure of real property rights is essential if we are to establish clearly defined procedures, to avoid the various mistakes made during the land reform and to carry out real property transactions in compliance with the main objective stated in the COST G9 action. In addition, parties that are involved in the process must be identified and their responsibilities clearly stated.

The development of infrastructure of real property rights comprises various activities, such as putting in place of appropriate legislative basis to protect personal privacy and public property, strengthening of societal background, informing and involving of society, improvement of professionals' skills, provision of the required material basis, maintenance of land registration and cadastral systems, and implementation of future development concepts. A systems approach is needed to carry out these measures, without emphasizing the details and concentrating on their interrelation.

Before preparing proposals or doing research into these particular future developments as mentioned above, one reasonable way would be to consider the impact of the societal background on land tenure by first describing and analysing the national variety of forms of land tenure according to the main transaction types, and by describing these transaction types.

Societal background

The societal background comprises an environment that has an impact on land tenure, and it is mentioned here for the purpose of illuminating the factors that influence land tenure in Latvia. This environment is related to the macroeconomic situation, political conditions, land policy and the implementation of the land reform, and existence of appropriate legislation; it also involves society, authorities and professionals, the real property market, traditions and other factors.

There are several socio-economic indicators that reflect the macroeconomic situation and have an impact on the forms of land tenure, i.e. overall unemployment and price levels and attraction of foreign investments to the sphere of construction.

Political conditions can be characterised by political stability, political and economic risk, and processes of integration.

Land policy reflects land management and administration issues, and they influence land tenure to a greater extent than other components of the societal background. For instance, implementation of land policy gives experience, which necessitates changes in legislation, involves society into various important processes, determines and influences the role of public administration and professionals, and the system of decision-making.

The ongoing activities for the purpose of sustaining land tenure are supported by legislation: the Civil Code of the Republic of Latvia and special laws enforced during the transition period. As far as cases are connected with restitution of property rights to former owners and their heirs, preference is given to specially drafted laws. Special legislative acts regulate the land reform and privatisation. Different provisions govern the land reform in urban and in rural areas.

Society, i.e. the people, participate in the ongoing processes by following rules and obeying decisions, and they are supposed to be subjects of real property rights and obligations as well as real estate users or possessors while the land tenure is established.

The role of public administration is divided between municipalities and various state institutions. Professionals dealing with land-related matters of technical, legal and economic nature are surveyors, planners, developers, notaries public, lawyers and property appraisers.

While there is a tendency to convert communal land tenure into individual tenure – a change from communal to individual stewardship of land, it is reasonable to observe the traditions that have an impact on the forms of land tenure.

Influence on land tenure

The influence of societal background on land tenure is seen through activities of persons having appropriate rights in real property. The purpose is to identify the various problems and tendencies: what are the reasons why during the 10 years of the land reform in Latvia all the land has not been transferred in ownership or at least legal possession rights granted?

The Republic of Latvia is a new country. Although there have been several land reforms that influenced the reorganisation of socio-economic conditions in the territory of Latvia at different times, the experience from the past has not been sufficiently taken into account. Just as land reformers in earlier centuries learned little from the reforms that had preceded them, the reforms of the twentieth century will, very largely, not achieve what is expected of them (Powelson, 1987).

Admittedly, at different times there were specific conditions, such as political ideology, rate of national economy development, aims of the land reform and other conditions, and in this light we may say that the emerging difficulties were more or less of an objective nature.

If we knew why appropriate problems arose during previous land reforms, then we might understand why similar failures are in process today (Powelson, 1987). The purpose of this contribution is not to analyse the past problems, but reflect on the consequences of the established societal background influencing land tenure in the time period between 1990 and the present time.

An overview regarding socio-economic conditions of Latvian situation gives Table 3 according to the data of the Central Statistical Bureau of the Republic of Latvia.

Table 3: Basic socio-economic indicators

Indicator / Year	1995	1996	1997	1998	1999	2000	2001
Unemployment rate (at the end of period), %	6.6	7.2	7.0	9.2	9.1	7.8	7.7
Consumer price changes, % of previous period	25.0	17.6	8.4	4.7	2.4	2.6	2.5
Foreign direct investment (flows),[16] million. lats	-	210.6	303.4	209.8	202.7	247.4	214.7
Construction cost changes, % of previous period	40.0	8.0	7.9	11.0	4.4	-1.9	-5.0

Source: Central Statistical Bureau, 2002 (http://www.csb.lv)

Economic development data, the main economic trends and indicators shown in Economic Structure in 2000 can be found in the materials of the Commission of the European Communities – 2001, Regular Report on Latvia's Progress Towards Accession (CEC Regular report, 2001).

The information mentioned in both sources table 3 and Regular Report on Latvia's Progress Towards Accession can give an idea about the possibilities and needs of people to be granted real property rights. Different people may have different rights and obligations for the same piece of land; the rights may be more valuable than the obligations are costly, or vice versa (Powelson, 1987). Applying this statement to the latest ten-year period of economic development of Latvia that influenced land tenure, we may say that in general people make choices according to the ratio of yield (benefit obtainable from dwellings, business, environment, etc.) to costs (tax, development costs, operating costs, etc). The scope of problems can be expressed in a nutshell: real property management for the purpose of development versus real property trade. One of the stimulating points for people is a hopeful prospect for further economic development.

Political stability in the country is also a matter that influences the strengthening of land tenure. More or less actively, society keeps track of events on the political scene. Politicians may have different views on the development of the country. Liberals primarily support rapid growth of economy, but leftists will concentrate on wellbeing of the nation and gradual changes. The Saeima (Parliament of the Republic of Latvia) as legislator and the government as executive power have the obligation, through negotiations and decision- making to achieve a way that would increase trustworthiness and stability. The Republic of Latvia is going to join the EU and NATO in the foreseeable future; therefore a lot of issues, including real property legislation, have to be addressed in conformity standards. The people on the whole prefer to live in conditions with diminished political risk that is closely linked with economic risk.

The system of land management in Latvia is rather decentralised. Municipalities perform public administration functions with regard to land management. Land management is a 'burden on the shoulders' of local

---

[16] Recalculated according to the exchange rate of the lats (LVL) for period average.

municipalities in Latvia, whereas various state institutions, i.e. the State Land Service of the Republic of Latvia (SLS), work in close contact with municipalities and implement land administration.

All matters related to cadastre and cadastral valuation of real estate are in competence of the State Land Service as laid down in Regulations of the Cabinet of Ministers of the Republic of Latvia ('Regulations on the State Cadastre of Real Estate'). The State Cadastre of Real Estate consists of both textual and graphical parts. The textual part of the cadastre includes data on real estate location, cadastral identification of land plots, and their area, information about buildings and structures, property value, encumbrances, easements, as well as information about owners or users. The graphical part of the cadastre contains digital cartographic material with boundaries of land plots and buildings, cadastral identification and other data characterising real estate. The SLS has 8 regional offices with representatives in 27 cadastral offices at the district level.

According to the 'Land Book Law', land registration offices fulfil legal real property registration functions. Twenty-eight land registration offices belonging to district courts are responsible for corroboration of real property rights. The Land Book register consists of sections or folios. Each folio has its own identification and is divided into four parts in which information about real property, owners, mortgages, easements and encumbrances is recorded.

Local municipalities – 498 in rural areas and 77 in towns – take decisions regarding purposes and types of land use as well as the shape and size of land parcels in their administrative territories. They also monitor and control compliance with these decisions. The structures of local governments co-operate with design and planning offices and employ planners and architects. They plan their respective administrative territories by drawing up master plans and detailed plans. Municipalities receive background information for development and updating of their plans from the SLS. The municipal planning offices confirm the respective land value zoning developed by the SLS. Land value zones are determined according to the cadastral value of land. Cadastral value is approximated to the market value, because land values must be updated according to the analysis of real estate market data. Market analysis is made by the SLS. The SLS gives advice to the municipalities when there is a wish to change the purpose of land use as well as controls the demarcation of real property boundaries.

The Land Commission is a structural unit of the local government that is directly responsible for decision-making in matters of land assignment. In rural municipalities Land Commissions have completed their work: land has been distributed. The final decisions on restitution of former property rights and assignment of land in possession for payment are taken by the SLS. Land Commissions are still continuing their work in all towns (Auzins, 2001).

Decisions, physical planning, housing regulations and other regulations issued by local governments are of crucial importance for property owners and users in deciding the form of land tenure.

Besides municipalities and the SLS, some other state institutions, for example, the State Forest Service, as well as different private structures are involved –

individuals and legal persons that participate in the implementation of land policy and the land reform.

The Civil Code and regulations concerning the land reform and privatisation constitute the body of legislation that regulates the ongoing activities and influences land tenure. The former Civil Code of 1937 came into force again during the transition period, and now, except for some special cases, regulates restitution of former ownership. The renewal of the pre-war Civil Code was a political decision.

The Civil Code consists of four parts: family rights, inheritance rights, rights in things, and rights of obligations. All of them, with the exception of the first part, are related to activities that influence land tenure. These parts of the Civil Code are subdivided into chapters. The second part, inheritance rights, mainly deals with inheritance by law, inheritance by contract and testamentary inheritance. The third part, rights in things, mainly regulates possession, ownership, easements, encumbrances, mortgages and redemption rights. The fourth part, rights of obligations, deals mainly with legal transactions, contracts, grants, and lease and hire contracts.

The main legislative acts applicable to the land reform are the Decision 'On Agrarian Reform in the Republic of Latvia', Law 'On Land Commissions', Law 'On Land Reform in Rural Areas of the Republic of Latvia', Regulations 'On Implementation of the First Stage of the Land Reform in Rural Areas of the Republic of Latvia', Law 'On Land Use and Land Planning', Law 'On Land Reform in Urban Areas of the Republic of Latvia', Law 'On Land Reform in Rural Areas of the Republic of Latvia', 'Land Book Law', Law 'On Completion of the Land Reform in Rural Areas', Law 'On Completion of the Land Reform in Urban Areas'.

Processes of real estate privatisation mainly are supported by the following legislation: Law 'On Privatisation of Property Units Owned by the State and Municipalities', Law 'On Land Rights of the State and Municipalities and their Corroboration in the Land Book', and Law 'On Registration of Real Property in the Land Book'.

A peculiar feature of the land reform was that one general law or a united set of several interdependent rules did not regulate it. It was regulated by several legislative and regulatory acts that were spontaneously made. The harmonisation process was sometimes rather long; separate articles were amended, even several times, during the adaptation process. As a result, there were controversial provisions in various regulatory acts and even within one particular regulation (Boruks, 2001:268).

Numerous amendments were made to the land reform legislation on the basis of practical experience gathered while carrying out the reform. Gaps in legislation caused disputes and led to compromises regarding real property rights; therefore in many cases granting of ownership rights was delayed. Local Land Commissions, the Central Land Commission and courts were made responsible for solving disputes related to real property rights. No special land court has been established in Latvia. The Central Land Commission is responsible for the co-ordination of the land reform and specification of all rules for the purpose of the reform.

Up till the last stage of the land reform there was no special law that would regulate matters of divided real property. Leases are generally regulated by the Civil Code, but more specifically by regulations applicable to the land reform. The need to adopt specific regulations concerning leases comes from the practical experience gained during the implementation of the land reform.

Legislation does not prescribe an obligation for people to have a title to real property, but it states that legally valid real property transactions today are possible only when a land registration office has appropriately corroborated the property rights.

The development of the real estate market is generally not very uniform. For instance, there is quite an active market in the capital city of Riga and in another nearby town – Jurmala. Different situations can be observed in other towns, as the real estate market is less active in rural areas. For two reasons the real estate market is to a great extent deformed: first, the validity of privatisation vouchers for the purpose of real property privatisation has been extended more than one time; second, the real estate market sometimes fails to function properly, mostly because of gaps in legislation.

Traditions are also a constituent part of the societal background; they influence land tenure and are reflected in social culture. On the one hand, the legacy from Soviet times has been too strong, particularly at the beginning of the land reform. For example, former 'bosses' and their relatives who sometimes acted against the principles of land reform and common sense, held public administration positions at local municipalities. On the other hand, there was a widespread view that 'I do not want another man's land, but I am not going to give away a single foot of my own land'. This was a characteristic opinion expressed by former real property owners (who lost their property in 1940) and their heirs. Such prejudices caused irreparable mistakes and long-term consequences, particularly in the capital city of Latvia, Riga.

Legal provisions on property transactions

The way in which rights in land are held is called 'tenure' (ECE/HBP/96 Economic Commission for Europe, 1996) . In a recording system, data concerning land tenure play an essential role. The concept of land tenure can, in this context, be defined as 'the act, right, manner or term of holding a landed property' or as 'nature of legal estate in land'. If land tenure is related to the broad field of the land use, it is more than the 'man-land' relationship. In that connection, it can be defined as the institutionalised relationship of people involved in the use of land and the distribution of its products.

Land tenure

The range of forms of land tenure that exist in Latvia reflects the time period from 1990 up till now – the period of transition connected with the land reform, and some variations of the notion of all normal purposes may be observed.

Ownership reflects fixed tenure through land registration and is linked with a complex set of rights and obligations (Powelson, 1987). A title registration system exists in Latvia; therefore the legal consequences of any transaction or privatisation case, i.e. the right itself (title), are subject to registration. So the right itself together with the name of the rightful claimant and the object of that right with its restrictions and charges are registered. With this registration the title or right is created.

Leasehold is practised when the owner, that on many occasions is the State – without title on particular real property, is either unable or unwilling to manage the real property and through an agreement transfers possession rights to another person who will fulfil certain duties for an agreed period of time. The lessee has to revert the real property to the owner in the quality similar to that on the date of its reception (Boruks, 2001). According to the agreed duration, there may be short-term leases (for not more than 5 years) and long-term leases (lasting more than 5 years). Normally the law regulates both real property rights and leasehold rights. There is no special law regulating leases in Latvia, even in cases of divided real property.

Land use is a term practically used in rural areas according to the regulations enforced for the purpose of the land reform. Land use as a conditional form of land tenure is valid only during the transition period while there is no real ownership on particular land plot. Land is assigned for use by a decision taken by the local municipality, at the same time specifying the purpose of land use; such decision can serve as a sufficient basis for surveying of boundaries. Thus land can be assigned in use for ten various purposes and the conditions are binding both in cases of restitution of real property rights and privatisation of real estate. Land use is classified into four groups, and the classification is used for the purpose of property taxation.

Adjudgement is a document that can be referred to as both political and legal recognition of previously granted real property rights (up till 1940) and of assignment of land in possession for payment (privatisation)[17] applicable to owners of buildings. Local municipalities issue adjudgements, which serve the purpose of boundary surveying.

Legal Possesion occurs when a person may own land which is in the possession of another person and occupied by a third. Possession involves the ability to enjoy the use of the land and in some circumstances to exploit the products on or below its surface and implies the physical power to control an object. A possessor of land has the ability to make use of the land in some way or another. Possession may be legitimate or illegal. The legal possession of another person's land takes place

---

[17] Assignment of land in possession for payment – granting of land purchase rights during the land reform in Latvia.

through formal agreements such as leases or rental agreements that protect the rights of the true owner.

Legal Possession Rights are given when land boundaries are planned, measured (by accurate measurement methods in urban areas) and demarcated, the relevant public administration authorities make the final decision on either restitution of previous property rights or assignment of land in possession for payment. In the latter case the potential possessor will purchase the land by concluding an agreement with the State or the local municipality. Next step is a visit to a notary public and the land registration office. Thus in the stage between issuing of the final decision and the corroboration of real property rights in a land registration office as prescribed by the Law 'On Real Property Tax' and Regulations on the State Cadastre of Real Estate, the possessor has legal possession rights in Latvia. The previously mentioned statement has a conditional character and is valid during the transition period because in most cases there is no true owner. The State and municipality have exclusive rights to proceed with real properties in those cases even when they do not have real ownership rights (title).

Free state land is a term used similarly to land use as a conditional form of land tenure and is valid exclusively during the transition period when there is no real property owner. The free state land means that land property rights have not been re-established to anyone; the right to use land has not been granted to anyone; land has not been assigned in possession for payment. Local municipalities manage the free state land, but up till now there are no regulations that would allow municipalities to own this free state land. Individuals and legal persons can claim such land and privatise it in the prescribed procedure.

Types of transactions

Generally transactions are regulated by the Civil Code and are specified accordingly in legislation applicable to the land reform. Article 1403 of the Civil Code states: 'Legal transaction is the action that is done in sanctioned order for the purpose of establishment, amendment or termination of legal relations'.

The land reform legislation (different for urban and rural areas) regulates transactions with land. In accordance with these provisions, transactions with land are any dealings that result in a change of land ownership, including inheritance by contract (testamentary inheritance), compulsory sale of mortgaged land and investment in corpora of a company of limited liability. There are rules that regulate the subjects of transactions, restrictions on transactions with land, the procedure of examination of transactions and other issues.

Conveyance of title is a normal case of land transactions when the real property rights, documented in particular title, are transferred from the owner to another person.

Conveyance of title includes transfer of ownership by selling-buying, granting or inheritance and can refer to every kind of divided real property that has a title.

Types of inheritance are described and procedures and restrictions determined by the Civil Code. Both inheritance by contract or testamentary inheritance are regarded as transactions. Inheritance by law is not considered a transaction. The procedure of inheritance is the responsibility of the court.

Compulsory acquisition for the purpose of public interests and the needs of the State is admissible in exceptional cases, and only for compensation. The compensation for expropriated real estate is granted in money. Law 'On Compulsory Acquisition of Real Estate for the Needs of the State or Public Interests' regulates compulsory acquisition.

Mortgage is a written instrument that creates a lien upon real estate as security for the payment of a specified debt. Mortgages are established by agreement, testament or through a court procedure. In the case of a mortgage the pledge is still in possession of the mortgagor, but the right of pledge is recorded in the land registry. If the owner (mortgagor) breaches provisions of the agreement, for instance, fails to settle the due payment on the fixed date, the mortgagee has right to honour the bill by taking over the pledge in ownership (Boruks, 2001).

Subdivision of land takes place in cases when the real property owner wants to sell lawfully part of his land to another person. An agreement must be concluded in these cases and the land subdivided in the cadastral procedure in accordance with the existing physical plans or detailed plans of the relevant administrative territory as well as in compliance with regulations issued by the municipality.

Investment in corpora of a company of limited liability is participation of an associate of Ltd company in actions of the company with an investment that can be real property. It is a specific kind of conveyance with regard to a corporation rather than to an individual.

Land tenure relation to the transaction types

Transactions are possible exclusively with land to which property rights have been corroborated in the land registry. It also means that only the true owner – the person that have a title is allowed to transfer particular land ownership rights to another party.

A distinction has been made between buildings and flats and landed property accordingly to the provisional Law 'On Real Property Registration in the Land Book'. It was allowed to make transactions with unregistered buildings until January 1st, 2000 (with the exception of buildings under privatisation, where transactions were allowed even one year later) and flat properties until January 1st, 2001. In such cases (selling-buying, grant, inheritance) the deal was based on a transaction contract concluded between parties and confirmed by the notary public, the responsibility for the registrations and other obligations were assumed by the property receiver. After the above-mentioned deadlines the transaction contracts involving buildings and flats are no longer valid without corroboration of property rights in land registration offices. As regards transactions with buildings, the rights of land use were transferred to the property receiver in cases when the property transferor had such rights. The land could be privatised some time later. The above-mentioned provisions refer to cases known as private conveyance.

At the same time, Article 994 of the Civil Code states: 'As real estate owner is recognised only such who has been as such recorded in land registries'.

Transactions with unregistered buildings and flats led to a lot of unlawful transactions and one of the main reasons why transactions of such kind were made legally possible, was the wish to promote the real estate market.

In Latvia rights to building properties can be sold, mortgaged, etc. separately from land ownership rights. It comes from a provision in the Civil Code that says: 'The one who has constructed a building on another person's land due to justifiable fallacy may ignore the claim to the land by the land owner until recompense is received for that building'. This statement was used to justify existence of property rights to buildings (Boruks, 2001).

In cases of transactions made by owners of building properties, the landowners have pre-emption rights, and vice versa. During the period of transition the interrelations between divided properties is generally regulated by the Civil Code, but is insufficiently specified by legislation on the land reform.

It is possible to mortgage unregistered real estate and get a loan from a commercial bank. Naturally, it is a great risk, but it is up to banks to grant or not to grant the loan. Frequently, when the pledge is a flat, the only source of reliable information is the cadastre.

Mortgaging of unregistered real estate was one of the factors that caused the banking crisis in the middle of 90s in Latvia.

The land reform legislation stipulates that it is possible to inherit land that has been assigned in use. This is applicable in cases when, before his death, the testator had the right to use land and to transfer land use rights to other persons. Heirs can apply for such a land in the set procedure.

Conclusions

The current land reform legislation provides that the land reform may be considered completed when all the land in Latvia has been legally assigned (in legal possession). It also means that all the land properties must be surveyed to measure their area. Legal possession does not mean that all the possessors have real ownership rights. Taking into account these conditions, a serious analysis must be made to find out what preconditions are required to accelerate the land reform and to complete it successfully, and to establish land tenure and real ownership through registration of real property rights in land registration offices.

The full implementation of the land reform requires that a rapid pace and good quality of implementation be ensured. In the near future it has to be decided whether to amend legislation, namely, regulations on the land reform and privatisation and the Civil Code, or to draft new legislation, for example, a Law 'On Transactions with Land'. That would help to normalise relationships between the currently independent types of real estate – land and buildings – in our country, because we have to start amalgamating these independent properties. Besides, in the foreseeable future some legislation must be provided for possible land consolidation projects.

At this stage of the ongoing activities one thing is clear: assignment of land in use, which is practised according to the Law 'On Land Use and Land Planning', has to be abolished. A new law that would regulate leasehold relations separately in both urban and rural areas must be made and enforced.

Anyway, in the political sense the land reform has reached two significant objectives. First, the restitution of ownership to the former owners or their heirs has been a significant investment in the renewal of the Republic of Latvia. Second, the process demonstrated a political will to enlarge and strengthen the role of private land ownership (Auzins, 2001).

A Law 'On Transactions with Land' should be adopted to achieve the following:

- to determine how cadastral (real property formation) procedures and transactions with real property may be done, taking into account that divided real property exists in our country (land and building on it can be independent objects of property rights);
- to support transactions with appropriately specified legislation – either the Civil Code or rules and regulations of the land reform or one separate law;
- to help to overcome the consequences of the divided land reform in urban and rural areas, to eliminate the existence of separate building and land properties;
- to promote the use of uniform real estate terminology and notions.

Provisions to be included in the Law have to be helpful not only in cases of land property formation for the purpose of transactions, but also where, due to mistakes previously made, real property documents do not conform to the situation in the field. Solutions for such situations are not given in any regulatory or legislative acts.

Development and maintenance of infrastructure of real property largely depend on the following:

- implementation of the ongoing processes with consideration of the future long-term perspective;
- initiation and strengthening of the role of public administration in land management and administration;
- strengthening of land management and administration systems, possibly centralizing through territorial reform;
- involving of society in the processes, i.e. focusing on future development concepts.

References

Auzins, A., (2001) Carrying on Land Reform in Latvia. International Conference 'On Land

Management and Administration', AGROBALT'2001, Vilnius 2nd-4th May 2001.

Boruks, A., (2001) Zemes izmantosana un kadastrs Latvija. LLU SZC, VZD, Riga, pp. 83-87.

CEC Regular report, http://europa.eu.int/comm/enlargement/report2001/lv_en.pdf

COST 328/00 (2001) http://cost.cordis.lu/src/action_detail.cfm?action=g9

ECE/HBP/96, (1996) *Land Administration Guidelines – with Special Reference to Countries in Transition.* UN, New York and Geneva.

http://www.csb.lv (1 February 2002).

Powelson, J.P., (1987) *The Story of Land. A world history of land tenure and agrarian reform.* Lincoln Institute of Land Policy. USA., ix, x, p. 308.

PART III
ONTOLOGICAL MODELLING

Chapter 7

Modelling Real Estate Transactions: The Potential Role of Ontologies

Ubbo Visser and Christoph Schlieder

Abstract

Ontologies have been identified as valuable formal models that support communication exchange. In this paper we introduce various and recent modelling methods and address their potential for modelling transactions in land in general. We identify drawbacks and argue that ontologies can help to overcome the obstacles. After discussing formal ontologies in more detail we conclude with specific domain ontologies and discuss the advantages and disadvantages.

Introduction

There is a wide range of methods and tools a system engineer can choose from in order to model transactions in land. We focus on two widely used methods, a modelling technique from software engineering (UML) and on an emerging method that is often referred to as ontological modelling.

Unified Modelling Language (UML)

The Unified Modelling Language is the industry-standard language for specifying, visualizing, constructing, and documenting the artefacts of software systems (Booch, Rumbaugh and Jacobson, 1999). It is built to simplify the complex process of software engineering. UML supports the process of system engineering, giving the user separate but closely linked graphical notations to represent object-oriented concepts.

It consists of graphical representations for class diagrams, object diagrams, collaboration diagrams, and sequence diagrams. Class diagrams represent object-oriented concepts and their attributes and operations. Object diagrams represent objects and their links. They can be seen as a snapshot of a running program. Collaboration diagrams are extensions of object diagrams by messages. It is possible to represent sequences of actions by giving hierarchical numeration, labelling operations, and giving directions of messages. Sequence diagrams represent the timing between messages of classes and objects.

It is important to note that UML supports the developer of a system for both the static structure of the system to be constructed and the dynamic behavior of the system. The dynamic behavior of the system defines states and modifications of objects over time. It also defines the communication between objects needed for certain services. Therefore, UML can play an important role with regard to the modelling of processes in general. Please refer to Sumrada in this chapter for further details about UML.

Ontological modelling

The term 'ontology' has been used in many ways and across different communities. In the following we will introduce ontologies as an explication of some shared vocabulary or conceptualisation of a specific subject matter. Ontologies have set out to overcome the problem of implicit and hidden knowledge by making the conceptualisation of a domain (e.g. geography) explicit. Provided that ontologies are encoded in a suitable language that can be processed automatically by a computer, ontologies can beneficially be applied in the following areas:

- *Systems Engineering:* The use of ontologies for the description of information and systems has many benefits. The ontology can be used to identify requirements as well as inconsistencies in a chosen design. It can help to acquire or search for available information. Once a systems component has been implemented its specification can be used for maintenance and extension purposes.
- *Information Integration:* An important application area of ontologies is the integration of existing systems. The ability to exchange information at run time, also known as interoperability, is an important topic. In order to enable machines to understand each other we also have to explicate the vocabulary of each system in terms of an ontology.
- *Information Retrieval:* Common information-retrieval techniques either rely on a specific encoding of available information (e.g. fixed classification codes) or simple full-text analysis. Both approaches suffer from severe shortcomings. Using an ontology in order to explicate the vocabulary can help to overcome some of these problems. When used for the description of available information as well as for query formulation, an ontology serves as a common basis for matching queries against potential results on a semantic level.

The above-mentioned benefits of ontologies for information modelling, exchange, and search give rise to their potential role in the context of modelling real property transactions. This role ranges across all mentioned benefits from systems engineering to information exchange.

Upper-level ontologies can be seen as theories that capture the most common concepts, which are relevant for many of the tasks involving knowledge extraction, representation, and reasoning (Kiryakov, Simov and Dimitrov, 2001). These ontologies are used to represent the skeleton of the human common sense in such a formal way that covers as many aspects of the knowledge as possible. A chair may serve as an example. Most of the chairs have four legs. However, there are exceptions, e.g. chairs with three legs or office chairs, which are sometimes considered to have one leg. What should be the cardinality of the attribute leg (given an object-oriented representation)? What is the minimal and maximal cardinality?

Therefore, most of the upper-level ontologies define their concepts loosely and mainly in taxonomic relations. One example for an upper-level ontology is the 'Upper Cyc Ontology' (Cycorp, 1997) with approximately 3.000 terms capturing the most general concepts of human consensus reality. Another prominent example is SENSUS (Knight and Luc, 1994), a 70.000-node terminology taxonomy. SENSUS is a framework into which additional knowledge can be placed. It is an extension and reorganization of WordNet (Fellbaum, 1998). The authors also added nodes from the Penman Upper Model (Bateman, Kasper, Moore and Whitney, 1990) at the top level, and the major branches of Word Net have been rearranged to fit.

Using upper-level ontologies for practical problems is not easy due to technical problems such as different representations and terminologies. In addition, there are no formal mappings between the upper-level ontologies available.

We have introduced two 'extreme' modelling approaches. Modelling with UML is well known and has advantages because UML supports both static knowledge and dynamic behaviour. A major disadvantage of UML-based modelling, however, is the non-existence of model checking, i.e. consistency checking. It is also not possible to make implicit knowledge explicit. The latter is the main advantage of formal ontologies. If written down in a logic-based language, consistency-checking and explicit construction of hidden knowledge with the help of inference mechanisms is possible. On the other hand, describing processes, e.g. workflow events, is not possible.

In short, both approaches have advantages, which can be used for modelling land transactions. A combination of both ideas would be ideal. However, the differences between the two approaches are too big. Therefore, we have to consider an intermediate approach to overcome the obstacles.

Ontologies and description logics

In the early 90s a new area around the idea of ontologies began to emerge. Gruber (1993) describes an ontology as a 'formal and explicit specification of a conceptualization'. This view of ontologies is widely accepted within the IT community. Leading researchers in the area claim that the above definition best characterizes the essence of an ontology (Fensel, van Harmelen, Horrocks, McGuinness and Patel-Schneider, 2001). A conceptualization refers to an abstract model of how people commonly think about a real thing in the world, e.g. a chair. Explicit specification means that the concepts and relations of the abstract model have been given explicit names and definitions. Formal means that the definition of terms is written down in a formal language with well-understood properties. Very often, a logic-based language is used for this purpose. It is important to note that the main thought behind the usage of this kind of language is the avoidance of ambiguities of concepts.

Grüninger and Uschold (2002) correctly argue that there are many kinds of things that people call ontologies. Following descriptions of concepts from one extreme to another we may have loose terms (less meaning) only (Figure 7). Following the line to the right the degree of meaning increases. The other extreme are descriptions of terms within formalized logical theories. Moving from left to right also means that the ambiguity of terms decreases.

Specifying ontologies

The descriptions of terms in ontologies are formal as mentioned above. One can argue that the description of classes and objects represented in UML is also formal. While this is true, there is a difference with regard to the degree of formalization.

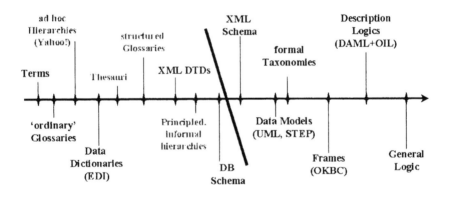

Figure 7: Types of ontologies (Grüninger and Uschold, 2002), the right hand side of the arbitrary line indicates what most researchers refer to as ontologies

We can differentiate between informal, semi-formal, and formal languages. The English language is an example for an informal language. Some terms are not well defined and it is easy to create ambiguities of concepts (e.g. 'spatial boundary'). Semi-formal languages are created to support software engineers developing software systems. UML is such a language but it is still open for ambiguities. Formal languages have a higher degree of formality. However, this does not imply that all these languages are eligible for our purpose. First-order logic for example is a formal language but is undecidable in general. Also, the language does not contain a model-theoretical semantics, which we need for reasoning support. Most of the description logics however support formal semantics and efficient reasoning support (e.g. OIL, see next section).

In order to demonstrate the differences in the above mentioned semi-formal and formal languages we give an example. As a use case we model a small part of the German ATKIS catalogue. ATKIS (Amtliches Topographisch-Kartographisches Informationssystem) is an official information system in Germany (AdV, 1998). It is a project of the head surveying offices of all German states. The working group offers digital landscape models with detailed documentation in the object catalogue OK-1000. This catalogue is the basis for our description. The example consists of seven classes in the areas Traffic and Waters. We are able to develop the following class diagram with UML:

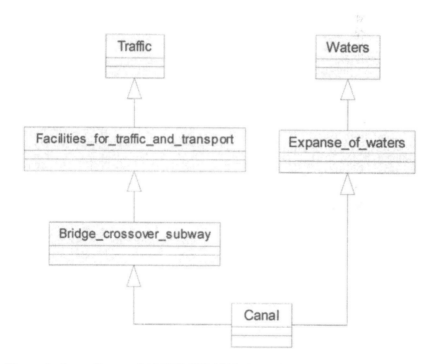

Figure 8: A small part of ATKIS OK-1000 classes in an UML class diagram

The catalogue describes Traffic and Waters as disjunctive object types. Figure 8 shows a small part of the dependencies between the classes. A canal is described as a type of object that is a crossover for transport. This crossover belongs to the super class Facilities for traffic and transport, which itself is a subclass of the main class Traffic. In addition, a canal is also a subclass of Expanse of waters, which itself belongs to the super class Waters. We can see that the class canal has two super classes. If Traffic and Waters are disjoint concepts this would lead to a problem in a running program. This example might be too easy and obvious. However, if we consider the vast amount of classes (the ATKIS OK-1000 catalogue e.g. has almost 100 classes) for a real world scenario one can imagine that the obvious problems shown in our example are not at all obvious in a real model. The developer can hardly overview the numerous classes and their meaning and the interactions between them. Therefore, he is not able to notice the conflicts in the first place.

Description logics

Description logics (DL) describe knowledge in terms of concepts and restrictions on roles. They also can derive classification taxonomies automatically. The main idea behind DL is to provide means to describe structured knowledge in a way that we can access and reason with it. DLs in general represent a class of logic-based knowledge representation languages. These languages are parts of first-order logic, which are both expressive enough and decidable with regard to inference mechanisms (Nebel, 1996). DLs are also known as terminological logics (Baader et al., 1991). Classical examples of description logics that are implemented are KL-ONE, LOOM, or CLASSIC.[18] A specific feature of DL is that classes (concepts) can be described intensionally by properties. These properties must be fulfilled for an object to belong to this class. Modern DL with efficient reasoning systems are SHIQ with the reasoner FaCT (Fast Classification of Terminologies, (Horrocks, 1999)) and RACER (Reasoner for A-Boxes and Concept Expressions Renamed, (Haarslev and Möller, 2001)). The example mentioned above can be described in DL notation as shown in Figure 9:

$$
\begin{aligned}
C &\sqsubseteq A \\
E &\sqsubseteq C \\
D &\sqsubseteq B \\
F &\sqsubseteq E \wedge F \sqsubseteq D \\
A &\parallel B
\end{aligned}
$$

with
A = Traffic
B = Waters
C = Facilities_for_transport
D = Expanse_of_waters
E = Bridge_crossover_subway
F = Canal

Figure 9: Example in DL notation

Over the last few years an initiative for the development of a language for the so-called Semantic Web (Bernes-Lee, Hendler and Lassila, 2001) has been

---

[18] http://www.research.att.com/sw/tools/classic/imp-systems.html

established. One result is the Ontology Interchange Language OIL (Fensel et al., 2000). OIL is partly based on DL from which it inherits its formal semantics. The other two roots are frame-based systems and Web standards such as XML and RDF. Fensel et al. (2001) claim that efficient work with ontologies requires support from advanced tools. There is a need for a language to express and to represent ontologies that meets three basic requirements:

- Intuitive to the human user (basically an object-oriented look and feel).
- It must have well-defined formal semantics with reasoning support.
- It must have a proper link with existing web languages.

OIL matches these criteria and is therefore a language of first choice. In order to demonstrate the power of this language with the supporting reasoning tool FaCT we modelled the same example as before. The classes and subclasses are shown in Figure 9. On the first level, the concepts Traffic and Waters are defined. At this point we also can define that traffic and waters are disjoint. At the second and third level we can model the subclasses accordingly. An interesting part can be seen at level four where we model that the concept Canal inherits its properties from both the super class Expanse_of_waters and Bridge_crossover_subway. We extracted all properties of the existing classes and all other classes for better reading.

We now have several options with this ontology. If we use the FaCT reasoner we can verify the model in the sense that we check its logical consistency. We also can query the system, e.g. which relationship to the concepts Canal and Waters have? Another option is to describe a real object and ask the system under which concept this object would be classified.

An important feature is the possibility to check the consistency of an ontology. This is done by a subsumption algorithm. If we check the ontology as described in Figure 9 the FaCT reasoner terminates with a failure. It detected an inconsistency based on the concept Canal. The verification fails because we decided on level one that the concepts Traffic and Waters are disjoint. Therefore, a concept on level four cannot be a subclass of both traffic and water.

This minimal example only shows one feature of the computational power of description logics. One can imagine that this feature can be of tremendous help while modelling a vast amount of concepts. Consistency checking is not possible with an UML-based modelling approach. This is a definite advantage of ontology-based modelling approaches.

```
ontology-definitions

//************ FIRST LEVEL ***************************
...
class-def defined Traffic
class-def defined Waters
disjoint Traffic Waters

//************ SECOND LEVEL ************************
...
class-def defined Facilities_for_traffic_and_transport
subclass-of Traffic

class-def defined Expanse_of_waters
subclass-of Waters
...
//************ THIRD LEVEL *************************
...
class-def defined Bridge_crossover_subway
subclass-of Facilities_for_traffic_and_transport
...
//************ FOURTH LEVEL *************************
...
class-def defined Canal
subclass-of
  Expanse_of_waters
  Bridge_crossover_subway

end-ontology
```

Figure 10: The same ATKIS classes modelled with OIL

So far we have seen what modern ontology languages and reasoner can perform on a terminological level. If we want to model the real world we have to consider not only classes but objects (instances, individuals) and the relations between them. Also, we have to consider concrete domain concepts which state concrete predicate restrictions for attribute fillers. An example would be to restrict the number of floors for a skyscraper to at least eight. At present, FaCT fails to compute concrete-domains. However, the above-mentioned reasoner RACER supports this type of concepts. The following example might give an impression about the idea.

The example is based on the German ALKIS catalogue for object types (AdV, 1999). It consists of two concept definitions, a building and a skyscraper with some of their properties. Among the buildings on the campus of the University of Bremen three individuals are modeled: the TZI-building, the MZH-building, and the GW2-building. The concept definitions can be seen in Figure 11. A building has at least one floor and the height is greater than 3m. It has a function and the building has a certain type (e.g. free standing, block). In addition, it has a type of use (sometimes unused). Also, the second concept skyscraper is a building with a restriction on the number of floors (at least eight) and the height (greater than 50m).

```
(equivalent Building
    (and (>=    Number_of_floors        1)
         (>     Height_of_building      3.0)
         (some Function_of_building    Function)
         (some Type_of_building        Type)
         (some is_used_for             Type_of_use)))

(equivalent Skyscraper
    (and (>=    Number_of_floors        8)
         (>     Height_of_building      50.0)
         (some Function_of_building    Function)
         (some Type_of_building        Type)
         (some is_used_for             Type_of_use)))

(implies Skyscraper Building)
```

Figure 11: T-Box definitions of the building example

The definitions of the real objects (instances of classes) can be seen in Figure 12. We see that the TZI headquarter building has three floors, its height is 22m, and its function is for research and education. If we present this description to the reasoner, the TZI headquarter building would (not surprisingly) be classified as a building. The second description (MZH-building) meets the requirements of a skyscraper. It will also be classified as a building because a skyscraper is also a building. This means that the reasoner can make hidden knowledge explicit. This is an important feature of terminological reasoning.

```
(instance TZI_HQ
   (and  (= Number_of_floors          3)
         (= Height_of_building        22.0)
         (some Function_of_building   Research_and_education)
         (some Type_of_building       Free_standing_building)
         (some is_used_for            Public)))

(instance MZH-Building
   (and  (= Number_of_floors          8)
         (= Height_of_building        91.0
         (some Function_of_building   Research_and_education)
         (some Type_of_building       Block)
         (some is_used_for            Public)))

(instance GW2-Building Skyscraper
   (and  (= Number_of_floors          4)
         (= Height_of_building        28.0)
         (some Function_of_building   Research_and_education)
         (some Type_of_building       Block)
         (some is_used_for            Public)))
```

Figure 12: A-Box of the building example

Since the building GW2 was modelled as a skyscraper but only consists of four levels and a height of 28m, a conflict would be detected. This is equivalent to the example with FaCT mentioned above, however, RACER is able to calculate with concrete domains, which is not very often supported with logical reasoning machines.

These few examples might give a little insight in what we can expect when modelling ontologies. We have seen that there are sophisticated languages available to express the relationships between classes, their objects and their attributes. There are also powerful reasoning machines available that are able to support the modelling process with ontologies in various ways.

Lately, supporting ontology tools have been developed. Fensel et al. (2001) mention

- Ontology editors: their purpose is to create ontologies. Examples are OilEd (Bechhofer, Horrocks, Goble and Stevens, 2001), OntoEdit (Sure et al., 2002), and Protégé (Noy et al., 2001)
- Ontology-based annotation tools are used to reference unstructured or semi-structured information sources with ontologies (e.g. XML DTD, XML Schema, RDF and RDF(S)). See also Annotea (Kahan and Koivunen, 2001)and CREAM (Handschuh, Staab and Maedche, 2001)

- Reasoning with ontologies, e.g. with the help of FaCT or RACER.

We have seen the benefits and drawbacks when modelling with ontologies. In the next section we describe approaches, which deal with domain specific ontologies with regard to real property transactions.

Domain-specific ontologies

Among others, models of real property transactions have two main areas where ontologies can be applied: (a) ontologies in the development of legal knowledge systems, and (b) ontologies for representing and reasoning about spatial objects.

Legal ontologies

In the area of legal reasoning there is a growing interest in so-called legal ontologies. These are explicit mostly formalized models of legal issues. Some ontologies have been proposed to support the formalization of legal argumentation. Early proposals like McCarthy's language for legal discourse (LLD) (McCarty, 1989) are rather specialized logics for encoding and reasoning about lines of argumentation in law. For this purpose special modalities like permitted, forbidden, obligatory, and enabled were developed. Stamper (1991) also emphasizes agents and activities as central notions in legal reasoning and introduces modelling primitives for both. In the mid nineties, work in the area of legal ontologies changed from defining logical languages towards the definition of conceptual models of legal knowledge that try to define building blocks of legal reasoning in a more comprehensive way. Examples of such conceptual models are Valente's functional Ontology of Law (Valente, 1995) and van Kralingen's Conceptual Ontology (Kralingen, 1995). Visser and Bench-Capon (1998) discuss the mentioned ontologies and present criteria for a comparison of the ontologies and discuss the strengths and weaknesses of the ontologies in relation to these criteria. One of their conclusion is that none of the ontologies seem to have adequate provisions to specify legal procedures.

The conceptual nature of these ontologies enables us to re-use them for our purposes. We took van Kralingen's Conceptual Ontology and encoded it in a web-based state-of-the art ontology language. The ontology encodes three main concepts of legal reasoning, namely norms, acts, and legal modalities. The intuitive meaning of these concepts is that a norm controls and restricts acts in terms of a legal modality. Thereby acts are complex descriptions of actions. Their definition includes the performing agent, his aim, intention and means as well as a description of the spatial and temporal context the act is performed in. In the special case of modelling property transactions, the spatio-temporal context of an act is especially important, because the area that is addressed by a norm has to be exactly specified and the time point of a certain act (e.g. the entry of an agreement into the register) strongly affects the validity of other legal acts (e.g. selling the same piece of land). Therefore there is a strong need to include aspects of space and time into an ontology used to describe real property transactions.

Spatio-temporal Ontologies

During the past decade, researchers from AI, Geoinformatics, Computational Linguistics, and Cognitive Science have made a joint effort to come to a better understanding of the ontological issues involved in the processing of spatio-temporal information. As a result, a variety of specialized calculi for representing spatial and temporal facts as well as for reasoning with them have been developed. Well-known examples are the Interval Calculus (Allen, 1983) which is mainly used in temporal reasoning and the Region Connection Calculus (Egenhofer and Franzosa, 1991; Randell, Cui and Cohn, 1992) that allows to draw inferences based on topological information. Such calculi constitute ontologies in the sense of Gruber (1993) because they provide an explicit specification of spatial relation concepts and they satisfy also Guarino's (1998) stricter definition of ontologies. In the area of Qualitative Spatial Reasoning (see (Cohn, 1997) for a survey), spatial relational calculi have been studied in connection with their computational properties. Frequently, the analysis of the algorithmic complexity of the reasoning problem gives hints for finding efficient heuristic reasoning strategies. This makes qualitative spatial relation calculi especially attractive for ontological engineering.

Much less effort has been devoted to the development of a top-level spatio-temporal ontology. However, an agreement has been reached about the basic design choices at the top-level. Galton (2001) summarizes the discussion by distinguishing ontologies of space based on:

- *Tessellation models vs. vector models*: This distinction relates to a technical issue in GIS, the representation of geographical data at the geometrical level. In tessellation models, the primary object of interest is a set of locations specified by a tessellation. Raster GIS implement tessellation models. Vector models focus on objects that are characterized by attributes and locations specified by coordinates (see Laurini and Thompson (1992).
- *Field-models vs. object-models*: By abstracting from the representational level, a distinction at the functional level of a GIS is obtained which reflects the previous distinction at the representational level. Field-models associate attributes with locations and are therefore related to tessellation models. Object-models associate attribute and locations with objects in a way that can be represented by vector model.
- *Continuous vs. discrete space:* Mathematical examples for continuous and discrete values are real and integers, respectively. While it is not difficult to capture this distinction ontologically by an appropriate axiomatization, there exist only approximations of the continuum on a digital computer. However, representations can be potentially continuous in the sense that they allow arbitrary further interpolation.

- *Absolute vs. relational space:* Space can be thought to exist independently of the objects existing in space. Absolute space commits to this ontological priority, which is inherent, for instance, to Newtonian physics. The opposite, relational space, claims that spatial and temporal entities (e.g. regions of space, time intervals) must be defined in terms of objects and their relations.

Although some of the choices in the above design alternatives are consistent with any other choice, there are two ways in which they are frequently combined. We call constellation ontologies those resulting from combining vector and object models with discrete representations and a relational view of space. In constellation ontologies, objects and relations are primary, whereas the space itself comes as a derived concept. Container ontologies, on the other hand, are those based on tessellation models and field-models with (an approximation of) continuous representations and an absolute view of space. In these top-level ontologies, space is primary in the sense that it constitutes a container for the objects.

A container ontology, the core ontology proposed by Coenen and Visser (1998), might serve as starting point of the iterative engineering process in which the spatio-temporal ontology for the European COST Action 'Modelling Real Property Transactions' is built. We propose to use this ontology because it can easily express approaches for qualitative spatial reasoning and because it also captures time. The ontology allows us to specify spatial objects, which can be used to define the spatial aspect of an act.

Conclusion

Modelling real property transactions is not a trivial task. We have to model static knowledge (e.g. parcels, buildings etc.). We also have to deal with processes, and we have to deal with abstract entities such as rights. We have seen that a logic-based formal language, which is supported by tools, has advantages when modelling static knowledge, and can therefore be used for this problem. In addition, there are already domain ontologies available, both in legal and spatio-temporal domains. However, we mentioned that modelling with ontologies has shortcomings. Their inability to describe processes might be one reason why they have not been used frequently in this kind of domains. UML on the other hand has advantages in exactly this area. Future work has to be done with regard to the inclusion of (legal) procedures within ontologies.

References

AdV. (1998). *Amtliches Topographisch-Kartographisches Informationssystem ATKIS.* Bonn: Landesvermessungsamt NRW, 801 pages.

AdV. (1999). *Amtliches Liegenschaftskataster-Informationssystem* (Technical report): Arbeitsgemeinschaft der Vermessungsverwaltungen der Länder der BRD.

Allen, J. F. (1983). Maintaining Knowledge about Temporal Intervals. *Communications of the ACM, 26*(11), 832-843.

Baader, F., Heinsohn, H.-J., Hollunder, B., Müller, J., Nebel, B., Nutt, W., and Profitlich, H.-J. (1991). *Terminological knowledge representation: A proposal for a terminological logic* (Technical Memo TM-90-04): Deutsches Forschungszentrum für Künstliche Intelligenz GmbH (DFKI).

Bateman, J. A., Kasper, R. T., Moore, J. D., and Whitney, R. A. (1990). *A general organization of knowledge for natural language processing: the PENMAN upper model* (Technical report). Marina del Rey, California: USC/Information Sciences Institute.

Bechhofer, S., Horrocks, I., Goble, C. and Stevens, R. (2001). *OilEd: A Reason-able Ontology Editor for the Semantic Web.* In: Proceedings of the KI 2001, Wien.

Bernes-Lee, T., Hendler, J., and Lassila, O. (2001). The Semantic Web. *Scientific American, 2001*(5).

Booch, G., Rumbaugh, J., and Jacobson, I. (1999). *The unified modelling language user guide.* Reading Mass.: Addison-Wesley, 482 pages.

Coenen, F. P., and Visser, P. (1998). *A General Ontology for Spatial Reasoning.* In: Proceedings of the ES'98, pp.44-57.

Cohn, A. (1997). *Qualitative spatial representation and reasoning techniques.* In: Proceedings of the KI-97: Advances in Artificial Intelligence, Berlin, pp.1-30.

Cycorp. (1997). Upper Cyc Ontology: Cycorp, Inc.

Egenhofer, M., and Franzosa, R. (1991). Point-set topological spatial relations. *International Journal of Geographic Information Systems, 5,* 161-174.

Fellbaum, C. (1998). *WordNet - An Electronic Lexical Database*: MIT Press, 423 pages.

Fensel, D., Horrocks, I., Harmelen, F. V., Decker, S., Erdmann, M., and Klein, M. (2000). *OIL in a Nutshell.* In: Proceedings of the 12th International Conference on Knowledge Engineering and Knowledge Management EKAW 2000, Juan-les-Pins, France.

Fensel, D., van Harmelen, F., Horrocks, I., McGuinness, D. L., and Patel-Schneider, P. F. (2001). OIL: An ontology infrastructure for the semantic web. *IEEE Intelligent Systems, 16*(2), 38-44.

Galton, A. (2001). Space, time, and the representation of geographical reality. *Topoi, 20*(2), 173-187.

Gruber, T. (1993). A Translation Approach to Portable Ontology Specifications. *Knowledge Acquisition, 5*(2), 199-220.

Grüninger, M., and Uschold, M. (2002). Ontologies and Semantic Integration, *Software Agents for the Warfighter*: Institute for Human and Machine Cognition (IHMC), University of West Florida.

Guarino, N. (1998). *Formal Ontology and Information Systems.* In: Proceedings of the FOIS 98, Trento, Italy.

Haarslev, V., and Möller, R. (2001). *High Performance Reasoning with Very Large Knowledge Bases.* In: Proceedings of the International Joint Conferences on Artificial Intelligence (IJCAI), Seattle, WA, pp.161-166.

Handschuh, S., Staab, S. and Maedche, A. (2001). *CREAM - Creating relational metadata with a component-based, ontology-driven annotation framework.* In: Proceedings of the K-Cap 2001 - First International Conference on Knowledge Capture, Victoria, B.C., Canada.

Horrocks, I. (1999). FaCT and iFaCT. In Lambrix, P., Borgida, A. and Lenzerini, M., Möller, R. and Patel-Schneider, P. (Eds.), *Proceedings of the International Workshop on Description Logics (DL'99)* (pp. 133-135).

Kahan, J., and Koivunen, M.-R. (2001). *Annotea: an open RDF infrastructure for shared Web annotations.* In: Proceedings of the 10th International World Wide Web Conference (WWW10), Hong Kong, pp.623-632.

Kiryakov, A., Simov, K. I., and Dimitrov, M. (2001, July 30 - August 1, 2001). *OntoMap - the Guide to the Upper-Level*. In: Proceedings of the Proceedings of the International Semantic Web Working Symposium (SWWS), Stanford University, California, USA.

Knight, K., and Luc, S. (1994). *Building a Large Scale Knowledge Base for Machine Translation*. In: Proceedings of the American Association of Artificial Intelligence Conference (AAAI), Seattle WA.

Kralingen, R. W. v. (1995). *Frame-based Conceptual Models of Statute Law*. The Hague, NL: Kluwer Law International, 232 pages.

Laurini, R., and Thompson, D. (1992). *Fundamentals of Spatial Information Systems*. (Vol. 37): Academic Press, Harcourt Brace & Company, Publishers.

McCarty, L. T. (1989). *A Language for Legal Discourse - 1. Basic Features*. In: Proceedings of the Second international on conference on Artificial Intelligence And Law, New York, pp.180-189.

Nebel, B. (1996). Artificial intelligence: A computational perspective. In G. Brewka (Ed.), *Principles of Knowledge Representation*. Stanford: CSLI publications.

Noy, N. F., Sintek, M., Decker, S., Crubezy, M., Fergerson, R. W., and Musen, M. A. (2001). Creating Semantic Web Contents with Protege-2000. *IEEE Intelligent Systems, 16*(2), 60-71.

Randell, D., Cui, Z., and Cohn, A. (1992). *A spatial logic based on regions and connection*. In: Proceedings of the 3rd Int. Conf. on Knowledge representation and reasoning, pp.165-176.

Stamper, R. K. (1991). The role of semantics in legal expert systems and legal reasoning. *Ratio Juris, 4*(2), 219-244.

Sure, Y., Erdmann, M., Angele, J., Staab, S., Studer, R., and Wenke, D. (2002). *Collaborative ontology engineering for the semantic web*. In: Proceedings of the International Semantic Web Conference ISWC, Sardinia, Italy.

Valente, A. (1995). *Legal Knowledge Engineering: A Modelling Approach*. Amsterdam: IOS Press.

Visser, P. R. S., and Bench-Capon, T. J. M. (1998). A Comparison of Four Ontologies for the Design of Legal Knowledge Systems. *Artificial Intelligence and Law, 6*(1), 27-57.

Chapter 8

A Tool-Supported Methodology for Ontology-Based Knowledge Management

York Sure

Abstract

A methodology for introducing an ontology-based Knowledge Management (KM) solution into enterprises is described which extends and improves the CommonKADS methodology by introducing – among others – specific guidelines for developing and maintaining the respective ontology. Special emphasis is put on a stepwise construction and evaluation of the ontology. The methodology is supported by a tool, OntoKick, that supports ontology engineers in early stages, i.e. the kickoff phase of ontology development. Though the main focus of the methodology is to support the introduction of KM solutions into enterprises, it contains relevant aspects for ontology development in general and provides therefore helpful support for the development of ontologies for other domains like real estate.

Introduction

Knowledge Management (KM) has become an important success factor for enterprises in virtually all areas during the last decade – to name but a few, one might think of human resource management, enterprise organization and enterprise culture. Information technology (IT) plays a crucial role in knowledge management, e.g. by operationalizing knowledge management processes in daily life.

IT-supported KM solutions are built around some kind of organizational memory (Kuehn and Abecker 1997) that integrates informal, semiformal and formal knowledge in order to facilitate its access, sharing and reuse by members of the organization(s) for solving their individual or collective tasks (Dieng et al. 1999). In such a context, knowledge has to be modeled, appropriately structured and interlinked for supporting its flexible integration and its personalized presentation to the consumer. Ontologies have shown to be the right answer to these structuring and modelling problems by providing a formal conceptualization of a particular domain that is shared by a group of people in an organization (O'Leary 1998).

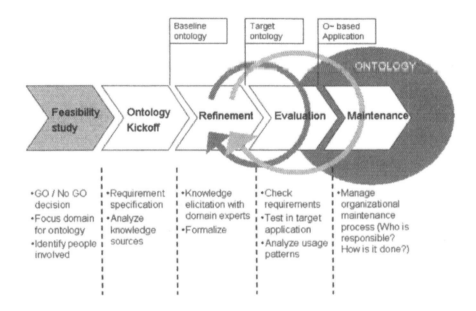

Figure 13: Steps of the methodology

Ontologies aim at capturing domain knowledge in a generic way and provide a commonly agreed understanding of a domain, which may be reused and shared across applications and groups. Ontologies typically consist of definitions of concepts, relations and axioms. Until a few years ago the building of ontologies was done in a rather ad hoc fashion. Meanwhile there have been some few, but seminal proposals for guiding the ontology development process (cf. Section on related work). One of the most prominent methodologies is CommonKADS that puts emphasis on an early feasibility study as well as on constructing several models that capture different kinds of knowledge needed for realizing a KM solution (Schreiber et al. 1999).

In this paper we first describe a methodology for introducing an ontology-based KM solution into enterprises which extends and improves the CommonKADS methodology by introducing – among others – specific guidelines for developing and maintaining the respective ontology. Special emphasis is put on a stepwise construction and evaluation of the ontology. Second, we present a tool, OntoKick, that supports ontology engineers in early stages of the methodology, i.e. the kickoff and the refinement phase of ontology development. Before we conclude we present related work. Though the main focus of the methodology is to support the introduction of KM solutions into enterprises, it contains relevant aspects for ontology development in general and provides therefore helpful support for the development of ontologies for other domains like real estate.

Methodology for Ontology-based Knowledge Management

In contrast to well-known methodologies for ontology development (cf. Section 4 on related work), which mostly restrict their attention within the ontology itself, our approach focuses on the application-driven development of ontologies during the introduction of ontology based knowledge management systems. We cover aspects from the early stages of setting up a knowledge management project to the final roll out of the ontology-based knowledge management system. The steps of our methodology are sketched in Figure 13, we will now describe each step in detail.

Feasibility study

Any knowledge management system may only function satisfactorily if it is properly integrated into the organization in which it is operational. Many factors other than technology determine success or failure of such a system. To analyze these factors, one must initially perform a feasibility study to first identify problem/opportunity areas and potential solutions, and second, to put them into a wider organizational perspective. The feasibility study serves as a decision support for economical and technical project feasibility, in order to select the most promising focus area, i.e. the domain for the ontology based system to be developed. We rely on the approach for carrying out a feasibility study that is described by the CommonKADS methodology (Schreiber et al. 1999). It should be carried out before actually developing ontologies and serves as a basis for the kickoff phase. Besides the domain of the system it helps to identify the people involved in setting up and using the system, viz. the domain experts, users and supporters of a system).

Kickoff phase

The first step to actually engineer ontologies is to capture requirements in an Ontology Requirements Specification Document ('ORSD') describing what an ontology should support and sketching the planned area of the ontology application. It should guide an ontology engineer to decide about inclusion, exclusion and the hierarchical structure of concepts in the ontology. In this early stage one should look for already developed and potentially reusable ontologies. In summary, it should clearly describe the information shown in Table 4.

Through analysis of the available knowledge sources a 'baseline ontology' is gathered, i.e. a draft version containing few but seminal elements of an ontology. Typically the most important concepts and relations are identified on an informal level. A very important knowledge source (also for the later phases) are domain experts. There exist several possibilities to capture knowledge from domain experts, we focus on the usage of informal competency questionnaires as proposed by Uschold and King (1995). They consist of possible queries (so-called competency questions) to the system, indicating the scope and content of the ontology. In the section on tool support for the kickoff phase of the methodology we will present a more detailed view on competency questionnaires and how we provide tool support based on their analysis.

Table 4: Content of the ORSD

ORSD1. Domain and goal of the ontology
ORSD2. Design guidelines to ensure a consistent development (e.g. naming conventions)
ORSD3. Available knowledge sources (e.g. domain experts, reusable ontologies, organization charts, business plans, dictionaries, index lists, db-schemas etc.)
ORSD4. Potential users and use cases
ORSD5. Applications supported by the ontology

Table 5: Two subphases of the refinement phase

R1. A knowledge elicitation process with domain experts based on the initial input from the kickoff phase. This serves as input for further expansion and refinement of the baseline ontology. Typically axioms are identified and modeled in this phase. This is closely linked to the next step – the effects of axioms might depend on the selection of the representation language.
R2. A formalization phase to transfer the ontology into the target ontology expressed in formal representation languages like DAML+OIL (DAML+OIL 2001). The representation language is chosen according to the specific requirements of the envisaged application.

This phase is closely linked to the evaluation phase. If the analysis of the ontology in the evaluation phase shows gaps or misconceptions, the ontology engineer takes these results as an input for the refinement phase. It might be necessary to perform several iterative steps.

Refinement phase

The goal of the refinement phase is to produce a mature and application-oriented 'target ontology' according to the specification given by the kickoff phase. This phase is divided into different subphases shown in Table 5.

Evaluation phase

The evaluation phase serves as a proof for the usefulness of developed ontologies and their associated software environment. In a first step, the ontology engineer checks whether the target ontology fulfils the ontology requirements specification document and whether the ontology supports or 'answers' the competency questions analysed in the kickoff phase of the project. In a second step, the ontology is tested in the target application environment. Feedback from beta users may be a valuable input for further refinement of the ontology.

A valuable input may be as well the usage patterns of the ontology. The prototype system has to track the ways users navigate or search for concepts and relations. With such an 'ontology log file analysis' we may trace what areas of the ontology are often 'used' and others which were not navigated. As mentioned before, this phase is closely linked to the refinement phase and an ontology engineer may need to perform several cycles until the target ontology reaches the envisaged level – the roll out of the target ontology embedded into the ontology-based application finishes the evaluation phase.

Maintenance phase

In the real world things are changing – and so do the specifications for ontologies. To reflect these changes ontologies have to be maintained frequently like other parts of software, too. We stress that the maintenance of ontologies is primarily an organizational process. There must be strict rules for the update-insert-delete processes within ontologies. Most important is to clarify who is responsible for maintenance and how it is performed. For example is a single person or a consortium responsible for the maintenance process? In which time intervals is the ontology maintained? We recommend that the ontology engineer gathers changes to the ontology and initiates the switch-over to a new version of the ontology after thoroughly testing possible effects to the application, viz. performing additional cyclic refinement and evaluation phases. Similar to the refinement phase, feedback from users may be a valuable input for identifying the changes needed. Maintenance should accompany ontologies as long as they are on duty.

Tool Support for the Kickoff Phase

Effectiveness and efficiency during the application of methodologies for system development increase significantly through tool support. Our tool OntoKick supports ontology engineers in the early stages of our methodology, viz. the kickoff phase of the ontology development. Tasks from the methodology are operationalized to enable e.g. up-to-date consistency checks and traceability of modeled objects like concepts and relations.

Speaking on a technical level, OntoKick is implemented as a plug-in (Handschuh et al. 2001) of the pre-existing ontology development environment OntoEdit (Sure et al. 2002), which allows for graphically oriented modelling of ontologies on a conceptual level (i.e. concepts, relations and to some extent also axioms are modelled independently from specific representation languages). The plug-in structure of OntoEdit enables flexible extensibility. Plug-ins can be uploaded during runtime and extend the range of OntoEdit's functionalities.

OntoKick captures stepwise the content of the Ontology Requirements Specification Document (ORSD) and builds a platform to integrate various elements. One might perform the steps ORSD1 to ORSD5 from Table 4 (cf. Number 1 in Figure 14) in the proposed order and perform cycles (Number 2). We now describe each step in detail.

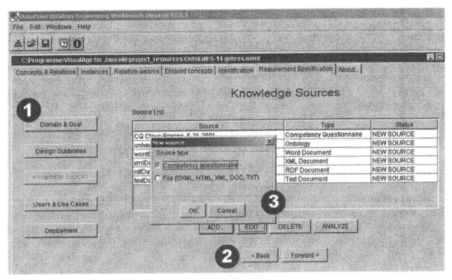

Figure 14: Management of different knowledge sources

Domain and goal are specified by ontology engineers at the very beginning of the development. To support reusability through the classification of ontologies we integrated standardized industry classifications.[19]

Design guidelines include 'predefined' guidelines for the size and the structure of ontologies like the estimated number of concepts per ontology, the estimated maximum depth of the concepts hierarchy and a free form for additional guidelines (like naming conventions). They all remain linked to ontologies and the predefines guidelines are constantly checked during the entire development process. Whenever the ontology exceeds predefined constraints, users are prompted – e.g. if the envisioned maximum number of concepts or the maximum depth of the hierarchy branch is exceeded. One might think of two possible reactions. Either

---

[19] We used two commonly known standardized industry classification for the implementation: (1) Industry Classifactions & Eligibility Requirements from Commercial News USA, http://209.208.147.42/CNUSA/US/sectors/ and (2) Hoover's Industry Sectors, http://www.hoovers.com/sector/0,2187,73,00.html

one needs to check the ontology itself to keep it within the boundaries. Or the guidelines, i.e. the requirements might need to be changed to reflect that they changed themselves (it seems natural that requirements might change over the time). So, the ontology is always within the range of specifications that constantly reflect the actually valid requirements.

Knowledge sources include all kinds of valuable knowledge sources for ontology development. OntoKick allows to manage different kinds of knowledge sources (see Number 3 in Figure 14) to keep the references to knowledge sources used during the development. For the analysis of the knowledge sources we currently focus on the capture of knowledge from domain experts with competency questionnaires, which will be described at the end of this section.

Users and use cases are specified similar to the knowledge sources, i.e. links to existing documents are stored to keep track of the used documents.

Finally, the deployment of ontologies to different applications is handled, i.e. each supported application might be characterized (e.g. the needed representation language for export of the ontology) and the path to store the productive version of the ontology might be specified.

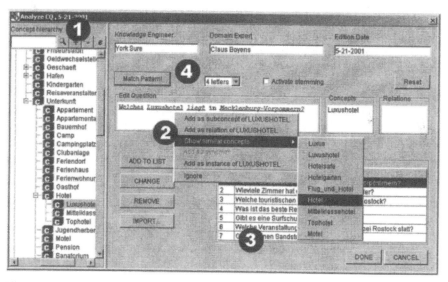

Figure 15: Competency questionnaire

Knowledge sources provide useful hints for the development of ontologies. Especially during the kickoff phase one needs support in finding relevant concepts and relations for the specified domain to gather a baseline ontology. They might be retrieved through analysis of knowledge sources. We identified competency questionnaires as one of the most important knowledge sources. Therefore we initially implemented a semi-automatic extraction of concepts and relations from competency questions (see Figure 15). For the future we plan to expand OntoKick to support the analysis of various kinds of knowledge sources like e.g. business documents, UML diagrams and mind maps.

An explorer-like tree shows the 'is-a' concept hierarchy of the actual ontology (Number 1 of Figure 15). As basic functionality from OntoEdit, concepts and their relations[20] can be added, deleted, renamed and restructured. Competency questions like "Welches Luxushotel liegt in Mecklenburg-Vorpommern" (Number 2), which means in English "Which luxury hotel is located in Mecklenburg-Vorpommern?", can be inserted and are stored in an enumeration of competency questions (Number 4) which altogether form a competency questionnaire filled in by a knowledge engineer and a domain expert.

From each competency question one may select elements (i.e. words) in order to define concepts, relations or instances[21] of concepts.[22] By default elements are inserted into the ontology as subconcepts of root (the uppermost element in the hierarchy), relations of root or instances of root. If a concept of the hierarchy is preselected, concepts / instances are added as subconcepts / instances of the marked concept and relations are attached to it.

During the development ontologies may grow big and finding relevant superconcepts gets time consuming. Therefore we implemented a simple pattern matching assuming that in some cases the subconcept contains the name of it's superconcept or at least parts of the name (Number 4). The ontology engineer specifies the number of letters that should match (four letters in our example). All elements of the competency question that have parts of the specified size (four in our example) that match to parts of that size of already modelled concepts are underlined. In Figure 15 for example, the words 'Welches', 'Luxushotel', 'liegt' and 'Mecklenburg-Vorpommern' are recognized by the system by using four letters for the pattern matching and, therefore, underlined.

A right click with a mouse offers a context menu to add the element to the ontology or to show similar concepts, i.e. the concepts matching. In our example four letters of 'Luxushotel' matched to four letters of 'Hotel' — either 'hote' or 'otel'. For this example choosing five letters would have done the job even better. In the figure shown we already included the element as a subconcept of 'Hotel' (and it is therefore shown in the hierarchy). Finally we implemented an algorithm for word stemming (Porter Stemming Algorithm[23]) that can be activated by a check-box (works only for English words). In combination with pattern matching it might help to decrease the number of words that match but are not helpful in finding relevant concepts.

---

[20] Please note that we have a 'frame-oriented' view on ontologies where relations are explicitly attached to 'their' concepts.

[21] Instances seem to be less important during the kickoff phase. In later stages one might want to have an initial set of instances to test and evaluate the ontology.

[22] If a domain lexicon is available in OntoEdit, one might also add an element as a synonym for specific lexical entries from that lexicon.

[23] The Porter Stemming Algorithm is available in Java at:
http://www.sbs.cs.olemiss.edu/345/stem6.html.

While browsing with OntoEdit through the ontology, it is possible to get the reference to the competency question that resulted in a specific concept (or relation etc.). This traceability helps to clarify the meaning of concepts, especially if more than one person are involved in the modelling of the ontology or if the ontology is reused by other persons than the creator.

Related Work

We here give an overview of existing methodologies for ontology development and show which of their ideas are adopted and expanded in our methodology.

Skeletal methodology

This methodology is based on the experience of building the Enterprise Ontology (Uschold and King 1995), which includes a set of ontologies for enterprise modelling. The guidelines for developing ontologies start with identifying the purpose of an ontology and then concentrate on the building of ontologies which is broken down into the steps ontology capture, coding, evaluation and documentation. A disadvantage of this methodology is that it does not precisely describe the techniques for performing the different activities. For example, it remains unclear, how the key concepts and relationships should be acquired. Only a very vague guideline is given.

We catch up the idea of competency questions and expand their usage. We not only propose to use them for evaluation of the system, but also for finding relevant lexical entries like concepts, relations etc.

KACTUS

The approach of Bernaras et al. (1996) was developed within the Esprit KACTUS project. One of the objectives of this project was to investigate the feasibility of knowledge reuse in complex technical systems and the role of ontologies to support it. The methodology recommends an application driven development of ontologies. So, every time an application is assembled, the ontology that represents the knowledge required for the application is built. Three steps have to be taken every time an ontology-based application is assembled:

The methodology offers very little detail and does not recommend particular techniques to support the development steps. Also, documentation, evaluation and maintenance processes are missing (Lopez 1999). In general we agree with the general idea of application driven ontology development and in particular with refinement and structuring, which is reflected by our proposal of the ontology development process.

Methontology

The Methontology framework from (Gomez-Perez 1996) includes:

- The identification of the ontology development process, which refers to which tasks (planning, control, specification, knowledge acquisition, conceptualization, integration, implementation, evaluation, documentation, configuration management) one should carry out, when building ontologies.
- The identification of stages through which an ontology passes during its lifetime.
- The steps to be taken to perform each activity, supporting techniques and evaluation steps.
- Setting up an ontology requirements specification document (ORSD) to capture requirements for an ontology similar to a software specification.

The methodology offers detailed support in development-oriented activities except formalization and maintenance and describes project management activities. We adopted the general idea of an ontology requirements specification document (ORSD), but modified and extended the presented version by our own needs.

Conclusions

We presented a methodology for introducing ontology based knowledge management into enterprises. Our methodology covers steps from early stages in knowledge management projects to the deployment and maintenance of an ontology based knowledge management system. The methodology already has been applied to case studies ranging from the implementation of the knowledge portal of our own institute (Stojanovic et al. 2001) to several industrial case studies (Staab et al. 2001; Sure and Studer 2001). Experiences gained while applying the methodology in further case studies will be integrated in future versions of our methodology.

Effectiveness and efficiency while performing steps of methodologies for system development increase through tool support. We presented tool support for early stages of our methodology, viz. the kickoff phase of ontology development, by our tool OntoKick. Two main aspects are covered: the capture of general requirement specifications for an ontology and the analysis of a specific requirement specification, viz. competency questionnaires. They serve as knowledge sources for the development of a baseline ontology. OntoKick enables semi-automatic extraction of concepts, relations and instances out of the competency questions. Traceability ensures that the context of extracted concepts, relations and instances is persistent.

For the future we plan to expand the analysis of knowledge sources (i.e. analysis of documents other than competency questionnaires like e.g. mind maps). A promising area for expansion seems to be the tighter integration of use cases (in UML diagrams). Furthermore we will expand the support in general for further steps of our methodology. For example, we plan to explore the possibility of supporting the usage of the captured and analyzed competency questions for the evaluation phase. The final goal is to have full fledged tool support (as far as possible) for the methodology for ontology based knowledge management.

Acknowledgements

We thank our colleagues and students at the Institute AIFB, University of Karlsruhe, without whom the research would not have been possible — especially Hans-Peter Schnurr (now ontoprise GmbH), Siggi Handschuh (AIFB) and Claus Boyens (AIFB). The research presented in this paper has been partially funded by EU in the IST-1999-10132 project On-To-Knowledge.

References

Bernaras, A., Laresgoiti, I., and Corera, J. (1996) Building and Reusing Ontologies for Electrical Network Applications. In Proceedings of the European Conference on Artificial Intelligence ECAI-96.

DAML+OIL (2001) http://www.daml.org/2001/03/daml+oil-index.

Dieng, R., Corby, O., Giboin, A., and Ribiere, M. (1999) Methods and tools for corporate knowledge management. *Int. Journal of Human-Computer Studies*, 51(3):567–598.

Gomez-Perez, A. (1996) A Framework to Verify Knowledge Sharing Technology. Expert Systems with Application, 11(4):519– 529.

Handschuh, S., Staab, S., and Maedche, A. (2001) CREAM – Creating relational metadata with a component-based, ontologydriven annotation framework. In K-Cap 2001 - First International Conference on Knowledge Capture, Oct. 21-23, 2001, Victoria, B.C., Canada.

Kuehn, O. and Abecker, A. (1997) Corporate memories for knowledge memories in industrial practice: Prospects and challenges. *Journal of Universal Computer Science*, 3(8).

Lopez, F. (1999) Overview of methodologies for building ontologies. In Proceedings of the IJCAI-99 Workshop on Ontologies and Problem-Solving Methods: Lessons Learned and Future Trends. CEUR Publications.

O'Leary, D. (1998) Using AI in knowledge management: Knowledge bases and ontologies. *IEEE Intelligent Systems*, 13(3):34– 39.

Schreiber, G., Akkermans, H., Anjewierden, A., de Hoog, R., Shadbolt, N., Van de Velde, W., and Wielinga, B. (1999) *Knowledge Engineering and Management – The CommonKADS Methodology*. The MIT Press, Cambridge, Massachusetts; London, England.

Staab, S., Schnurr, H.-P., Studer, R., and Sure, Y (2001) Knowledge processes and ontologies. *IEEE Intelligent Systems*, 16(1):26– 34.

Stojanovic, N., Maedche, A., Staab, S., Studer, R., and Sure, Y. (2001) SEAL – A Framework for Developing SEmantic PortALs. In K-Cap 2001 – First International Conference on Knowledge Capture, Oct. 21-23, 2001, Victoria, B.C., Canada.

Sure, Y., Erdmann, M., Angele, J., Staab, S., Studer, R., and Wenke, D. (2002) OntoEdit: Collaborative ontology development for the semantic web. In Proceedings of the International Semantic Web Conference 2002 (ISWC 2002), June 9-12 2002, Sardinia, Italia.

Sure, Y. and. Studer, R. On-To-Knowledge Methodology – Final Version. Institute AIFB, University of Karlsruhe, On-To-Knowledge Deliverable 18, 2002. http://www.aifb.uni-karlsruhe.de/WBS/ysu/publications/OTK-D18_v1-0.pdf

Uschold, M. and King, M. (1995) Towards a Methodology for Building Ontologies. In Workshop on Basic Ontological Issues in Knowledge Sharing, held in conjunction with IJCAI-95, Montreal, Canada.

Chapter 9

Building a Foundation for Ontologies of Organisations

Chris Partridge and Milena Stefanova

Abstract

This paper presents a report on work in progress of a project to build a foundation for ontologies of organisations. The first stage of which is to synthesise a base enterprise ontology from existing ontologies, which will be used as the foundation for the construction of a Core Enterprise Ontology (CEO). The synthesis is intended to harvest the insights from the selected ontologies, building upon their strengths and eliminating – as far as possible – their weaknesses. The current work focuses on organisation, and one of its main achievements is the development of the notion of a person (entities that can acquire rights and obligations) enabling the integration of a number of lower level concepts. In addition, we have already been able to identify some of the common 'mistakes' in current enterprise ontologies – and propose solutions.

Introduction

This paper results from a collaboration between two projects: the BRont (Business Reference Ontologies)[24] and European IKF (Intelligent Knowledge Fusion)[25] projects.

The BRont project is part of the BORO Program, which aims to build 'industrial strength' ontologies, that are intended to be suitable as a basis for facilitating, among other things, the semantic interoperability of enterprises' operational systems.

This European IKF project has as an ultimate goal the development of a Distributed Infrastructure and Services System (IKF Framework) with appropriate toolkits and techniques for supporting knowledge management activities. The following countries participate in the IKF project; Italy, UK, Portugal, Spain, Hungary and Rumania. The project will last 3.5 years, and started in April 2000.

---

[24] http://www.BOROProgram.org

[25] http://www3.eureka.be/Home/projectdb/PrjFormFrame.asp?pr_id=2235

There are a couple of vertical applications whose domain is the financial sector. One of these, IKF/LEX – a part of the Italian IKF project – has been selected to undertake a pilot project. IKF/IF-LEX is lead by ELSAG BankLab SpA and its goal is to provide semi-automatic support for the comparison of banking supervision regulations.

There will be two kinds of ontologies developed within the IKF project:

- A Reference Ontology composed of a Top Level Ontology and several Core Ontologies (Breuker et.al. 1997). The top level ontology contains primitive general concepts to be extended by lower-level ontologies. The core ontologies span the gap between various application domains and the tope level ontology. The IKF/IF-LEX and the BRont projects are collaborating on developing a Core Enterprise Ontology (CEO) that IKF will use on this and its other applications in the enterprise domain.
- Domain Ontologies. The vertical applications will build ontologies for their specific domains. For example, the IKF/IF-LEX project is building an ontology for bank supervision regulations, focusing on money laundering.

Synthesis stage work plan

The scope of the synthesis work is large – and so the work has been divided into more manageable chunks.

As Breuker and others (Breuker 1997) state, a core ontology contains 'the categories that define what a field is about.' A first rough intuitive guess of what these categories might be has proved a useful tool in:

- helping clarify the scope focus on the important aspects for the CEO
- acting as a basis for segmenting the work.

The selected categories are:

- parties (persons) which may enter in
- transactions (composed of agreements and their associated activities),
- assets.

The ontologies to be analysed were selected according to:

- the relevance of their content to the Core Enterprise categories, and
- the clarity of the characterisation of the intended interpretations of this content (Gruber 1993), (Partridge 1996) and (Guarino 1997).

This gave us the following list:

- TOronto Virtual Enterprise – TOVE (Fox et al. 1996),

- AIAI's Enterprise Ontology – EO (Uschold et. al. 1997),
- Cycorp's Cyc® Knowledge Base – CYC,
- W.H. Inmon's Data Model Resource Book – DMRB (Hay 1997), (Inmon 1997).

The work proceeds by analysing one category in one ontology at a time, and then re-interpreting the previous results in the light of any new insights. Initially, the work focuses on individual ontologies but as it proceeds there is enough information to start undertaking comparisons between ontologies. The final analysis will encompass analyses of both the individual ontologies and comparisons between them.

In each of the ontologies, the concepts and relations relating to the category being considered are examined for the clearness and uniformity of their descriptions and formalisations. Further, each concept is analysed for its coverage and extendibility in cases where the coverage is not complete. Relations between concepts that are not explicitly described, but clearly exist, are identified as well. In addition, for the sake of a clear interpretation, we have found it necessary to consider the top concepts (whether or not they are explicitly described).

An important part of the analysis is testing each concept and its relations against a number of standard examples and more specialized concepts. Further, a check is made against a number of standard difficult cases. Both these checks help to identify weaknesses in the coverage of the ontologies.

A key concern in the analysis is to understand how the various concepts interlink with one another, to better understand the unifying structure of the Enterprise ontology.

At various stages during the analysis an interim ontology is synthesised from the strengths found in the analysis, in such a way as to eliminate the known weaknesses – and itself analysed. In the final synthesis, all the categories in all the ontologies are combined into a base CEO ontology.

At this time, we are concluding the analysis of the Parties (Persons) category for the EO and TOVE ontologies – and early drafts of synthesised ontologies are being reviewed. There is still substantial work that needs to be done in determining the precise relations between concepts, such as LEGAL ENTITY and OWNERSHIP within the EO.

Initial findings

Though both the ontologies have many important insights and provide much useful material – our most general findings, at this stage, are that none of the ontologies:

- adequately meet our criteria of clear characterisation, or
- really share a common view of what an organisation is.

Taken together, these findings mean that the creation of the synthesised base CEO ontology cannot just be a simple merging of the common elements of the selected ontologies.

We now illustrate these findings with examples. We also show how we synthesised a resolution to some of these problems – for the two ontologies we have analysed.

Cloar Charactorisation

With an unclear characterisation it can be difficult to work out the intended interpretation – in the worst case, impossible to decide between competing interpretations. There are many different ways in which the characterisation can be unclear – as we show below.

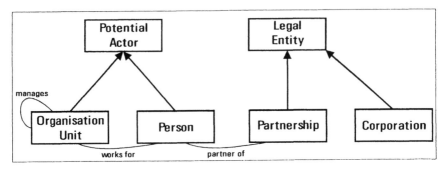

Figure 16: Simplified EO overview

In both TOVE and EO we found no clear overview of the structure – so we developed graphical representations based upon ER diagrams to help us understand it. Figure 16 and Figure 17 provide simplified versions of these.

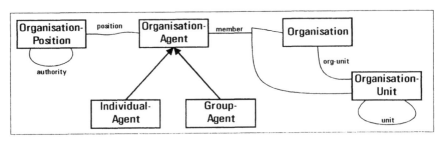

Figure 17: Simplified TOVE overview

Both TOVE and EO make use of a number of top concepts. A top ontology – or top concepts – can provide a useful structure for defining and using domain concepts and relations – segmenting the enterprise and other domains into general categories. However, if this is not done properly it can have the opposite effect.

Some of the problems we encountered with the top concepts and the domain analysis are:

- Insufficient characterisation of the disjointness of top concepts: For example, in the informal EO the relationship between the top concepts ENTITY, and ROLE is not clear – in particular, whether ROLES can be ENTITIES or not, and so whether they can enter into RELATIONSHIPS. The same lack of care in characterising disjointness (and overlapping) exists at the domain level in both TOVE and EO. We found this can make it impossible to definitely determine the intended interpretation. For example, in TOVE the formalisation allows an ORGANISATION-UNIT to be an ORGANISATION – though this seems counter-intuitive, and probably not what the authors intended.

- Not applying top concepts: TOVE states that a fluent is 'a [type of] predicate or function whose value may change with time'. But it does not identify which predicates in its ontology are fluents – leaving this to the readers, who have to make their own judgements. Supplying such information would have helped not only the users of the ontology but also its creators and designers. For example, the TOVE's creators end up (probably unintentionally) having to regard ORGANISATION as a fluent – when in the normal (commonsense) use of the concept it is not.

- Messy formalization trajectories: EO formalizes its concepts in logical systems (Ontolingua and KIF), which rely on their own (different) top concepts. An attempt for a clear formalisation trajectory has been made (Uschold et. al. 1997), but unfortunately this does not match very well with the informal specification. For example, in the informal EO it is stated that each RELATIONSHIP is also an ENTITY, but is not defined as such in the formalization. Furthermore some RELATIONSHIPS are defined in the formalization as classes and others are defined as relations without explaining what the motivations for these choices are (e.g., SALE is a RELATIONSHIP formalized as a class, HAVE-CAPABILITY is a RELATIONSHIP formalized as a relation). This becomes a more serious problem if the formalisation is meant to be taken as the more accurate version.

- Failing to use general concepts to achieve uniformity: Both TOVE and EO fail to use top concepts to describe in a uniform way core relations and concepts. This hampers understanding. Typical examples are the part-of relation, used in describing the decomposition of organizations into smaller units, and the relation, which shows the different ways for participation in organizations. For example, TOVE introduces two kinds of part-of relations: org- unit (between ORGANISATION and ORGANISATION-UNIT), and unit (between two ORGANISATION-UNITs). These relations express ORGANISATION and

ORGANISATION-UNIT decompositions, but are not explicitly unified under a common relation. In the EO several ways of participating in a company are considered, as a partner (partner_of relation between PERSON and PARTNERSHIP), as an employee (works_for relation between PERSON and OU), as a shareholder in a corporation (only in the informal EO specification, see Uschold et. al. (1997)). These ways of participation are not unified in the EO.

- Insufficient analysis. As an example consider the EO concepts of OWNERSHIP and SHAREHOLDING (Uschold et. al. 1997) which are formally unrelated, while SHAREHOLDING as evident from its informal and formal definitions represents the ownership relation between a CORPORATION and its owners.

Common view of an organization

Figure 16 and Figure 17 give a broad picture of the concepts included in the analysis of TOVE and EO. As even a cursory glance can tell there are significant differences.

There are many examples in both TOVE and EO of how a better analysis would have led to more similar views:

- Insufficient analysis: In TOVE, for example, it seems that an ORGANISATION is not an AGENT, but has AGENTS as members. Yet there are many examples of organisations (such as the EU or NATO), which have other organisations as members.
- Missing Links: In the EO, the relation between the concepts OU and LEGAL ENTITY is unclear. All that we are told is that a LEGAL ENTITY 'may correspond to a single OU' (Uschold et. al. 1997). No further analysis (informal or formal) of the link between these two concepts is given.
- Implicit context dependencies: In the EO, the concept LEGAL ENTITY, is not well thought out – having several (informally inconsistent) descriptions. It seems that the intended meaning actually depends on a particular jurisdiction (in this case on the current UK jurisdiction) – though it is not clear that the authors recognise this. This dependence is inappropriate in the modern global economy – and it raises potential problems should the UK jurisdiction change. For example, the LEGAL ENTITY concept would no longer be the 'union of PERSON, CORPORATION, and PARTNERSHIP'.

Unifying the core concepts: person

Part of the synthesis work is to analyse the ontologies in preparation for a synthesised common view. A vital missing element from both the ontologies is a unifying core category.

To resolve this, we have introduced the concept PERSON (PARTY), which can be a NATURAL PERSON or SOCIALLY CONSTRUCTED PERSON (SOCIAL

PERSON in short). This acts as the catalyst for transforming the ontologies into ones with similar characteristics. The next step (which we will undertake soon) is to merge them into a single synthesised ontology.

The result of introducing PERSON into the EO ontology is shown in Figure 18. A comparison of this with Figure 16 shows how PERSON has unified the taxonomy.

To give the reader some idea of how the transformation was effected, we describe the steps we went through. The EO concepts LEGAL ENTITY and OU are generalized into the concept PERSON. The EO concept PERSON (human being) is renamed into NATURAL PERSON. OU becomes SOCIAL PERSON, while LEGAL ENTITY is taken completely out and substituted with the context independent notion of LEGALLY CONSTRUCTED PERSON (LEGAL PERSON in short).

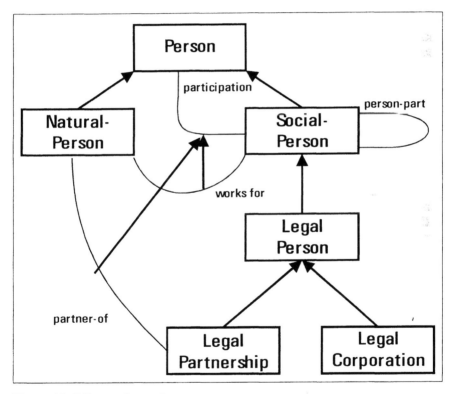

Figure 18: EO transformation

Note that LEGAL PERSON is not the same concept as the EO LEGAL ENTITY, since it is intended to represent parties which are constructed according to a legal jurisdiction, but not necessarily recognised by it as legal persons (in EO terms, LEGAL ENTITYs). For example, in UK a partnership is not legally recognized as a person (it cannot sign contracts in its name) but it is a LEGALLY CONSTRUCTED PERSON, because there are legal constitution rules for partnerships. Finally the two participation relations, partner-of and works-for are consolidated under a general participation relation, and the relation manages is renamed into person-part (which is a particular kind of part-of relation).

The result of introducing PERSON into the TOVE ontology is shown in Figure 19. As before, a comparison of this with Figure 17 shows how PERSON has unified the taxonomy. The transformation steps between Figure 17 and Figure 19 are similar in many respects to those between Figure 16 and Figure 18.

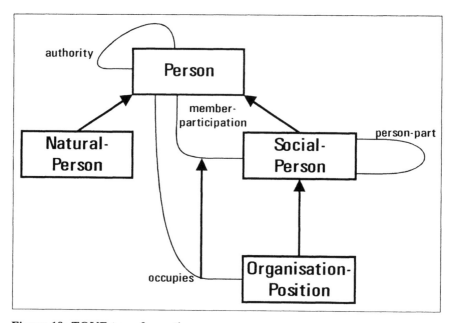

Figure 19: TOVE transformation

Conclusions

Even at this early stage our work has revealed the need for a substantial improvement in enterprise ontologies to bring them up to 'industrial strength'. Hopefully, our work will go some way towards realising this.

Acknowledgements

We would like to thank the IKF Project in general and ELSAG SpA in particular for making this research possible. Furthermore we would like to thank Alessandro

Oltramale, Claudio Masolo, and Nicola Guarino for the numerous fruitful discussions we had on topics related to ontologies and organisations.

References

Breuker J, Valente A. and Winkels R. (1997) Legal Ontologies: A Functional View in P.R.S. Visser and R.G.F. Winkels, *Proceedings of the First International Workshop on Legal Ontologies.*

Fox, M.S., Chionglo, J. and Fadel, F. (1993) A Common-Sense Model of the Enterprise, *Proceedings of the Industrial Engineering Research Conference.*

Fox, M.S., Barbuceanu, M. and Gruninger, M., (1996) An Organisation Ontology for Enterprise Modelling: Preliminary Concepts for Linking Structure and Behaviour, *Computers in Industry*, Vol. 29, pp. 123-134.

Gruber, T. (1993) Toward Principles for the Design of Ontologies Used for Knowledge Sharing, in Nicola Guarino and Roberto : *Formal Ontology in Conceptual Analysis and Knowledge Representation.*

Guarino, N. (1997) Semantic Matching: Formal Ontological Distinctions for Information Organization, Extraction, and Integration. In M. T. Pazienza (ed.) *Information Extraction: A Multidisciplinary Approach to an Emerging Information Technology.*

Hay, David C. (1997) Data Model Patterns: Conventions of Thought, Dorset House.

Inmon W.H. (1997) *The Data Model Resource Book: A Library of Logical Data and Data Warehouse Models*, John Wiley and Sons.

Partridge, C. (1996) Business Objects: Re-Engineering for Re-Use, Butterworth-Heinemann.

Uschold, M., King, M., Moralee, S. and Zorgios, Y. (1997) *The Enterprise Ontology*, AIAI, The University of Edinburgh.

Uschold, M., King, M., Moralee, S. and Zorgios, Y. (1998) The Enterprise Ontology, in *The Knowledge Engineering Review*, Vol. 13, eds. M. Uschold and A. Tate.

PART IV
SYSTEMS ENGINEERING

Chapter 10

Conceptual Modelling of Cadastral Information System Structures

Radoš Šumrada

Abstract

This paper presents a framework for the formal models of the cadastral system that is conceived as a land information management system. The objective is to describe the basic setting for modelling the inner and outer views of such systems, developing and using a specific (stereotyped) modelling methodology for conceptual formalism. Each view represents a certain aspect of the information system scrutinized. The internal view represents the inner static structure of the system, its data and therefore facts. The outer view is dynamic and stands for the users' or externally observed aspects of the system that shape its procedural and processing behavior or supplied services.

The analytical dissection of cadastral information system bases on the system analysis approach and relies on the object-oriented methodology. The modelling approach is based on the standardized geographical information developments, its reference models and terminology. The formal descriptive technique uses UML (Unified Modelling Language), which is the standardized conceptual schema (graphical and lexical) language. The combined glossary of the related and standardized terms is also added at the end. The modelling approach and the terminology used in the paper derive from two domains, which are GIS technology standardization and information or software systems modelling. The main sources are therefore CEN TC 287 prestandards and technical reports, the similar ISO TC 211 ones, and the OMG (Object Management Group) industrial standard UML.

Interpretation and models of reality

Reality is considered as infinite space and time, which we conceive as the complex actuality that surrounds us, and that as well permanently changes. Any description of reality is an abstraction that forms one of its possible interpretations. Therefore, our notion is an abstract model of the selected part of reality and as such never forms an entire representation. Selected features are approximated and simplified. Other instances, their relations and properties are ignored. The level of abstraction and selection is biased by the foreseen purpose and usage of the model (CEN ENV 12009:1998).

A model is applied abstract supplement that represents simplified mapping of reality into the conceived and interpreted notion, which manifests as a descriptive and graphical specification of the selected part of reality. A model captures the important aspects of reality from a certain point of view and simplifies or omits the rest. Which ones are important is a matter of judgment that depends on the purpose of the model. Modelling is the process of model developing and is generally based on a chosen methodology and practical constrains as well. Modelling is a well-proven and widely accepted engineering technique (Booch et al., 1999).

Models of a system

A system is a set of elements that are possibly arranged into subsystems, which are organized to accomplish a specific intention. Models of a system take on different forms for various purposes and appear at different levels of abstraction (Rumbaugh et al., 1999). A system is thus described by a set of models that describe it from different viewpoints. Models help us to understand, learn and shape both a problem domain and its solution domain. A model is a simplification of the selected part of reality that helps us to master a large and complex system, which cannot be comprehended easily in its entirety. The model is intended to be easier to use for certain purposes than the complete system observed. Models therefore unable us to organize, retrieve, examine and analyse data about large systems. Models as well evolve over time. Models with greater detail are derived from more abstract ones as the knowledge of the system expands over the development or maintenance process.

Spatial data and information

Firstly, the definition of data and information is defined and the difference between the two meanings is outlined. The described difference between the notion of data and information is respected further on in this text. Data are facts, ideas or instructions represented in a formalized manner, which is appropriate for communication, interpretation or processing by humans or computers. Information is the possible meaning of data, which humans derive from data through the known means of presentation and interpretation.

Spatial (geographic) data can be defined as computer treatable form of facts concerning phenomena directly or indirectly associated with a location relative to the Earth. Spatial data describes the thematic and cartographic characteristics and as well as various relationships among spatial phenomena, the location of which is geo-coded using a reference system. Spatial data are generally kept in a dataset that is an identifiable collection of geographic data, and which can further consist of several subsets.

Geographic data models serve as the foundation on which GIS (Geographic Information Systems) databases are formed. Spatial data represent gathered knowledge about the spatial phenomena concerned and include semantic, spatial and quality aspects. The semantic aspects describe the meaning of gathered facts and the characteristics of the derived model, or shortly the interpretation of data or some metadata. The spatial aspects define its position, geometry and topology. The quality aspects indicate its potential or fitness for the particular usage. Understanding defined geographic data model concept is central to know how to analyse and interpret geographic information.

Models of information system

An information system is a combination of database, human and technical resources that together with the appropriate organization means and personnel skills produce information needed to support certain economic activity, management of resources and decision-making procedures. We can interpret and model an information system from many varying viewpoints. The most important aspects are as follows:

- Data or database (static) models;
- Users' requirements models resolving system responsibilities;
- Business, transactions and organizational models;
- Application or processing (dynamic) models.

A model of a large information system permits us dealing with complexity that is difficult to handle directly (Rumbaugh et al., 1999). A model can be suitably abstract without getting lost in the details. The level depends on its purpose and must be comprehensible to humans. Models have semantic aspect and a visual presentation. Semantic elements carry the concepts or meaning of the model. The visualization shows the semantics in a graphical form (notation) that can be seen, browsed and edited. The following related issues, which are combined and interleaved, mainly conduct the development or a renewal of an information system.

Problem domain - The internal view

The domain is real, abstract or hypothetical field of endeavour under consideration, which can include various groups of objects that behave accordingly to the rules and characteristics of this domain (Rosenberg et al., 1999). The term problem domain refers to the area that encompasses real world features and concepts related to the problem that the information system is being designed to solve. Domain modelling is the tasks of discovering objects and their classification that represent those instances and concepts. The result is an abstract static model of reality, which we show on class diagrams and that is based on the selected semantics, formalism and terminology. The notion of 'problem domain' derives from information and software engineering and is similar to the terms 'universe of discourse' and somehow narrower term 'nominal ground' that are applied in the ISO and CEN geographic information standardization settings.

System requirements - the formal view

System requirements are an arrangement of things accountable and related together as a whole that represent the crucial responsibilities the system must manifest and fulfil (Rosenberg et al., 2001). This formal and mostly non-functional view of the system shows the general and common rules, such as standards, laws, regulations etc. that govern and affect the data supply process, formal procedures, the obligatory database content and services, which the information system should perform.

Users requirements - the external view

The essence of use case modelling is to capture all-important users' requirements of the new or renewed system by detailing all the scenarios that the users will perform. This dynamic model of an information system starts with the use case analysis that involves working inward from the user requirements. The result is a use case model, which is the external view of the system, forming the conceptual centre of such approach. This dynamic model also drives the static or data model. The result of use case modelling should be that all the required system functionality is described in the use cases.

Use case is a sequence of actions that an actor (a person, external entity or another system) performs within a system to achieve a particular goal (Fowler et al., 1999). A complete and unambiguous use case describes one aspect of usage of the system without presuming any specific design or implementation. An actor represents a role a user can play with regard to a system, or an entity such as database or another system, which resides outside the system being modeled. An actor can perform many uses cases and different actors can carry out a particular use case. The total set of actors within a use case model reflects everything that needs to exchange information with the system. We show use cases and actors as associations on use case diagrams that demonstrate the functional requirements the system should support.

The key for a use case is that each one indicates a function or a service that the user can master and that has certain value for that user. A use case can describe one or more paths through the user operations to accomplish the case and as well the system reactions. The basic course or scenario must always be present and some alternate scenarios are optional (Fowler et al., 1999). The basic scenario of a use case is the main start to finish course of action or the critical processing path the user will follow under normal circumstances in order to obtain the required service. An alternate course of action can be present as an infrequently used or parallel path through the exceptional scenario or error conditions.

Data and processing behaviour – the static and dynamic aspects

Any specific model of information system must first define the universe of discourse or the key concepts, their internal properties and relationships to each other. This static concept consists of classes, each of them describing a set of instances or discrete objects that hold the data and communicate by messages to invoke the processing behaviour. The data they hold are modelled as attributes or data members, and the behaviour they perform is modelled through their operations. Several classes can share their common structure through inheritance (generalization or specialization).

There are two ways to model the processing behaviour of classes forming the dynamic aspects of the system. The first approach is called the state machine and models one discrete object as it interacts with its environment. The other view is called an interaction and shows the communication pattern for a set of related objects as they collaborate in order to implement the common task or generally a use case. The view of a system through the interacting objects, their links and the flow of messages across data links can be interpreted in a sequential manner or as a collaboration and management pattern. This interaction models also serve as evolutional link between the analytical aspects (what), and the design (how) and implementation (where) views of the system (Rosenberg et al., 2001).

Unified modelling language (UML)

Unified Modelling Language (UML) is a general-purpose visual modelling language, which is used to specify, visualize and document the components of a discrete system (Booch et al., 1999). UML is an industrial standard (current version 1.4), which is under the supervision of the Object Management Group (OMG). It is important to realize that UML and the methods that use it are separate, but as well tightly related issues. UML is a modelling language and not a standard modelling method, but still it effectively replaces the past modelling practices.

The modelling language is the most important part of any development method and certainly is also the crucial one for communication. UML is intended to support most existing object-oriented analytical and design methods, and it enables an incremental and iterative development process. It consists mainly of the graphical notation that various methods can use to express analytical results or particular designs. The graphical notation is the basic syntax of this modelling language. Modelling a system from several separate but related viewpoints permits it to be understood for different purposes. The various views of a system can be graphically represented on the nine sorts of UML diagrams with the standard content, which can be extended and enhanced by additional user defined stereotypes.

UML captures the data about the static structure and the dynamic behaviour of a system, which is modelled as a collection of classified objects. They interact to perform services that ultimately benefit its users. The static structure defines the classes that are important for a system, their properties and relationships. In UML any large system can be further decomposed into subsystems that are represented as packages of related classes. The dynamic behaviour defines the states and modifications of objects over time and the needed communications among them in order to accomplish certain service. The starting point is use cases modelling that generally drive the whole development process.

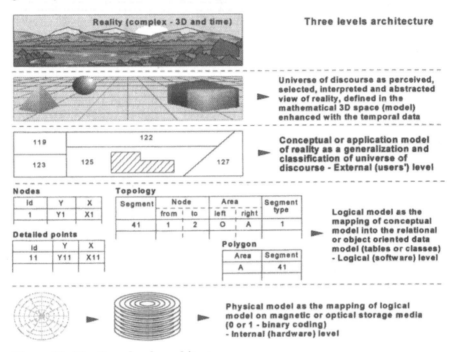

Figure 20: The three levels architecture

Conceptual modelling of geographical space

A formal model of space is an abstract and well-defined set of related concepts. Conceptual modelling is the process of creating an abstract description of some portion of the real world (ISO DIS 19101:2001). It follows so-called three levels architecture: external or conceptual, logical and internal views, as outlined on the Figure 20. Real word objects of the selected part of reality are mapped into the constructs that can be represented like classes in the information system database. Conceptual model defines concepts of a universe of discourse for a particular system and as such provides the abstract description of the selected real-world features. Conceptual formalism is a set of modelling concepts used to describe a conceptual model. The conceptual formalism provides the rules, constraints, inheritance mechanisms, events, functions, processes and other elements that make up a conceptual schema language (ISO DIS 19103:2001).

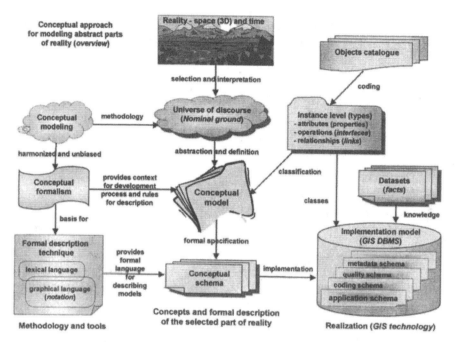

Figure 21: The conceptual modelling approach

Conceptual schema is formal description of a conceptual model in conceptual schema language for some universe of discourse (ISO DIS 19103:2001).

Conceptual schema language is the formal description technique used, which bases on a conceptual formalism for the purpose of representing conceptual schemas. A conceptual schema language provides the semantic and syntactic elements used to describe the conceptual model rigorously in order to convey its meaning consistently. A conceptual schema language may be lexical, graphical or both. Several conceptual schema languages can be based on the same conceptual formalism (UML, EXPRESS, IDEF1X etc.). The general overview of the whole conceptual modelling approach is presented on Figure 21.

Methodology for a cadastral information system development

UML is called a modelling language and is not a method as such (Fowler et al., 1999). Many modelling methods or conceptual formalisms consist of both a formal description technique and a development process. The description technique today is mainly a graphical language and notation (UML), which the applied method uses to express analytical results and implementation designs by visualizing them on various diagrams. The development process is the know how on what steps to carry out during analysis, design and implementation. The most methods today that incorporate UML are widely used in software engineering processes and as such are probably not directly transferable into the GIS technology domain.

Table 6: Main questions and solutions

Questions	UML techniques
1 Who are the users of the information system (the actors) and what they trying to do by interacting with the system?	Use case diagrams
2 What are the problem domain objects (real world abstractions), their classes and the relationships among them?	High level (analytical) class diagrams
3 What objects are needed for each (particular) use case?	Robustness analysis (class) diagrams
4 How do the objects participating within each use case interact?	Sequence and collaboration diagrams
5 How will be handled real-time control issues?	State and action diagrams
6 How are we going to design, develop and implement the system?	Detailed (design) class diagrams

Source: (Rosenberg et al., 1999).

As outlined above, UML is a general-purpose visual modelling language that is designed to be independent of any development methodology. All UML does is to say what the model diagrams may consist of and what they mean. The selection, abstraction, classification and specification are the liability of the developer. The crucial issue is therefore to develop or adopt a suitable method that is particularly tailored for conceptual modelling in the spatial data problem domain. The proposed approach is certainly based on object-oriented paradigm and follows the best practices from software engineering domain. However the stress should be on transparency and proper simplicity, because good methodology does not need to be complicated as well (Rosenberg et al., 1999).

The main stream of the proposed development process can be described by deriving detailed class models through use case models, which are the provenance of this approach and drive all the stages and procedures. Use case models capture user requirements for the information system by detailing all the scenarios users will be performing. This development process tries to answer to the questions about the information system deployment that are outlined in Table 1. The significant features of this development process are among the others as follows:

The approach is incremental and iterative. The static model (classes) gets refined incrementally during multiple successive iterations through the dynamic modelling, which bases on the use case analysis. The process works firstly inward from the users' requirements and system responsibilities, and later on outward from the data that is needed to support the processing services.

The approach offers a reasonable degree of traceability. Along the development steps the reference to the users' requirements is maintained and often also verified. The development trend proceeds from high-level analysis models to the detailed design ones.

The approach bases on the usage of UML standard as the formal description technique (conceptual schema language) and tries to apply the minimal set of required steps for the development process.

The basic steps through analysis and design that comprise this object oriented development approach of an information system are presented in the following summary (Rosenberg et al., 2001):

- Gather available data and expertise about the legacy of the renewing information system, similar existing systems, or user requirements for the new one. Do some simple rapid prototyping of the conceived and reengineered system.
- Identify the problem domain objects and their relations. Perform their classification (classes), generalization and aggregation in order to derive their basic structure. Specify their basic properties (attributes) and important relationships (associations and other dependencies) among them. Draw high-level analytical class diagrams that introduce the first cut data structure or the inner view of the system. This domain model serves also as a glossary of terms that are used during the whole use case modelling process.
- Identify and describe use cases, actors and the interactions between the actors and use cases (associations and other relationships). Develop use cases from the general to the very detailed ones. Present the use case model on a set of use case diagrams that mostly present the outside view of the system.
- If appropriate, structure and organize use cases into the groups. Present the organization of use cases on the package diagram.
- Requirements review represents a detailed sequential description of each use case. Write down a detailed basic course of action (critical path) and also all the alternate less frequent scenarios (alternatives, exceptions, error handling etc.).
- Perform robustness analysis, which means, for each use case identify the set of participating objects (instances of classes): (1) Identify all the objects that participate and accomplish each use case and its scenarios. (2) Try to factor out the similar processing behaviour or so-called patterns, which describe the common ways of performing tasks. (3) Update class diagrams with the newly discovered objects (classes), relationships and new attributes.
- Strive to foresee functional requirements for classes in order to support the realization of each use case.

- Finish updating the improved and detailed class diagram, which reflects the completion of the analytical phase of the development process.
- Allocate (processing) behaviour or methods to the classes and for each use case define all the interactions. (1) Identify the messages that need to be passed between objects and presume associated and invoked methods. Draw sequence diagrams to show and analyse message interchange. Update class diagrams with new discovered methods and attributes. (2) If needed, use collaboration diagrams to show key transactions between objects, which form the critical path or the main scenario of each use case. (3) If needed, use state diagrams to show real-time behaviour and dynamics of (important) objects.
- Finish the static model (class diagrams) and the dynamic model (interaction diagrams) by adding also more detailed design, non-functional and implementation requirements.
- Perform information system, its components and user acceptance testing by executing all the use cases.

Conclusions

According to the described methodology UML models for the simplified Slovene cadastral subdivision case have been developed. The outcomes are presented on the several UML diagrams.

References

Bennett, Simon, McRobb, Stephen and Farmer, Ray (1999), '*Object-oriented Systems Analysis and Design Using UML*'. McGraw-Hill Publishing Company.

Bennett, Simon, Skelton, John and Lunn, Ken (2001), '*Schaums Outlines: UML*'.

Booch, Grady, Jacobson, Ivar and Rumbaugh, James (1999), 'The Unified Modelling Language User Guide' Addison-Wesley Object Technology Series.

CEN ENV 12009 (1998), '*Geographic information - Reference model. (CEN/TC 287)*'.

Fowler, M. and Scott, F. (1999), '*UML Distilled: Applying the Standard Object Modelling Language*', (Second Edition). Addison Wesley.

ISO DIS 19101 (2001), '*Geographic information - Reference model. (ISO/TC 211)*'.

ISO DTS 19103 (2001), '*Geographic information - Conceptual schema language. ISO/TC 211*'.

Jacobson, Ivar, Booch, Grady and Rumbaugh, James (1999), '*The Unified Software Development Process*'. Addison-Wesley Object Technology Series.

Object Management Group (2000), '*OMG UML Specification 1.4.*' Technical Report. (see http://www.omg.org).

Rosenberg, Doug and Scott, Kendall (1999), '*Use case driven object modelling with UML: a practical approach*'. Addison-Wesley Object Technology Series.

Rosenberg, Doug and Scott, Kendall. (2001), '*Applying Use Case Driven Object Modelling with UML: An Annotated e-Commerce Example*'. Addison-Wesley Object Technology Series.

Rumbaugh, James, Jacobson, Ivar and Booch, Grady (1999), '*The Unified Modelling Language Reference Manual*'. Prentice Hall, Inc.

Chapter 11

Ontology Construction for Geographic Data Set Integration

Harry Uitermark

Abstract

In order to integrate different geographic data sets a conceptual framework is needed. In this chapter an ontology-based framework will be demonstrated. An ontology is a structured collection of unambiguously defined concepts. There are two kinds of ontologies: (1) ontologies for certain disciplines, domain ontologies, and (2) application ontologies, one for every data set involved. Other components of the framework are sets of surveying rules. Surveying rules determine the transformations from terrain situations into geographic data sets. With the help of application ontologies and surveying rules a top-level domain ontology for topographic mapping is refined and restructured into a reference model, in such a way that this reference model expresses the semantic interconnectedness of data sets. This chapter demonstrates how this refinement and restructuring is done.

Geographic Data Set Integration

Geographic data set integration (or map integration) is defined in this research as the process of establishing relationships between corresponding instances in different, autonomously produced, geographic data sets of the same geographic space (Uitermark 2001). Traditionally, in existing map series, corresponding instances were linked implicitly by a common spatial reference system, for example the national grid (Devogele et al 1996; Sester et al 1998; Kilpeläinen 2000). Geographic data set integration aims at making links between corresponding instances explicitly by investigating the way geographic data sets were acquired. Motivation and background of this research is update propagation, which is the reuse of updates, from one geographic data set into another geographic data set. Update propagation is studied within the range of traditional topographic data sets, or map series (van Wijngaarden et al 1997; Uitermark et al 1998; Kim 1999; Vogels 1999).

An Informal Introduction to Geographic Data Set Integration

The collection of geographic data in order to produce a paper map, is an activity that has been going on for centuries. Since the 1970s, geographic data is not stored on paper but in electronic, digital form. First in traditional plot files, and nowadays mostly in a dedicated information system with a special database, called a Geographic Information System, abbreviated as GIS. This availability of geographic data in digital form makes it relatively easy to combine geographic data sets of different origin, provided that these sets are of the same geographic space, and can be transformed to a common reference system (and therefore same scale). This transformation to a common reference system is sometimes trivial, or sometimes extremely complicated, see for example (Laurini 1998). However, after this transformation another problem pops up, if one wants to compare and interpret the combined data sets on the basis of individual data elements, and draw conclusions from these comparisons. This is the problem of geographic data set integration.

Take for example maps of two simple geographic data sets (Figure 22). Assume that both maps are from the same geographic space, and are the same 'snapshot' in time. They look similar although there are differences. How are we able to decide whether they resemble each other, or are different from each other? A simple overlaying of both maps might reveal coinciding areas. However, in order to interpret and draw valid conclusions from these coinciding areas, a necessary condition is the understanding of the semantics, the meaning of data sets.

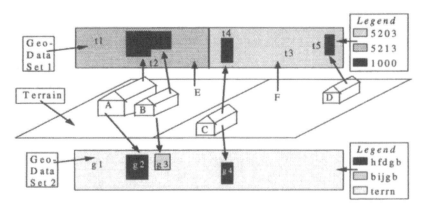

Figure 22: Two transformations of a terrain situation. In comparing geographic data sets it is mandatory to know surveying rules in order to conclude if data sets are consistent with the same terrain situation.

Inspecting the legends of both maps in Figure 22, the semantics of both data sets is far from clear. What are class labels as '5203' or 'hfdgb' supposed to mean? These class labels refer to classes with definitions within different data models. To reconcile these different data models, it is useful, even mandatory, to investigate the way geographic data sets were acquired, which is to say how the transformation was from real-world phenomena to data sets.

But then there is still a problem. In order to express and compare surveying rules, used in the acquisition of data sets, a collection of common ideas, or notions, of terrain objects is needed. This collection of common definitions of terrain objects is in many cases not available, because geographic data sets are produced independently by different organizations, all with their own objectives and ideas about terrain objects. Therefore it is necessary to invent or construct a collection of common definitions of terrain objects. Here is where a domain ontology is born, a collection of shared concepts, as a 'umbrella' for understanding object definitions in different geographic data sets.

To illustrate ideas as surveying rules and domain ontology take the simple terrain situation in Figure 22 (middle). There are four buildings, labelled A, B, C, and D, and two parcels E and F. In our domain ontology we have definitions for buildings and parcels, as well as their properties. This terrain situation is acquired with two different sets of surveying rules:

According to surveying rules of Geographic Data Set 1:

- buildings A, B, C, and D are acquired, and represented as t2, t4, and t5 with label '1000' in the map of Geographic Data Set 1 (Figure 1, above). Observe that A and B are merged into t2, because A and B are sufficiently close to each other. 'Sufficiently' has a precise definition in the surveying rules of Geographic Data Set 1
- parcel E (grass land) and F (arable land) are acquired, and represented as t1 (label '5213') and t3 (label '5203') in the map of Geographic Data Set 1 (Figure 22, above).

According to surveying rules of Geographic Data Set 2:

- buildings A, B, and C are acquired, recorded with different properties, and represented as g2 (label 'hfdgb'), g3 (label 'bijgb'), and g4 (label 'hfdgb') in the map of Geographic Data Set 2 (Figure 22, below). Building D is not represented because its area size is too small. Again, 'too small' is precisely defined in the surveying rules of Data Set 2
- parcels E and F are acquired, and represented as g1 (label 'terrn') in the map of Geographic Data Set 2 (Figure 22, below). E and F are merged into g1 because surveying rules state that the recording of different properties of E and F is not relevant for Geographic Data Set 2.

In order to understand semantic interconnectedness of both geographic data sets, domain ontology concepts such as 'building' and 'parcel' are refined into concepts as 'mainbuilding', 'annex next to mainbuilding', 'free standing annex', 'arable land', and 'grass land'. By structuring these concepts in a reference model, where concept labels refer to class labels, meaning is given to the hidden semantics of geographic data sets (Figure 23).

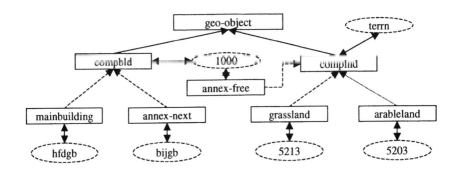

Figure 23: Refined concepts from a domain ontology are structured in a reference model (concept labels in rectangles). Concepts refer to class labels (in ovals), revealing semantic interconnectedness of geographic data sets.

With a reference model it is possible to reason or form hypotheses about terrain situations which are consistently represented in both data sets. This is what geographic data set integration is about. To do this reasoning, relationships between data elements from different sets, the corresponding instances, must be known, and these relationships must also be consistent with surveying rules of the data sets involved. Otherwise one can not determine whether the data sets in Figure 1 are consistent with the same terrain situation.

A first outcome of integrating the geographic data sets of the preceding example is a list of relationships between candidates for corresponding instances:

{ {(t2, g2), (t2, g3)}, {(t1, g1), (t3, g1), (t5, g1)}, {(t4, g4)} }

In a subsequent action, candidates for corresponding instances are checked for consistency with surveying rules.

From now on, if there is a modification in a terrain situation, which is succeedingly recorded for Geographic Data Set 1, it is clear from relationships between corresponding instances, if and how Geographic Data Set 2 will be influenced. This will be a starting point for update propagation.

A Conceptual Framework

This section provides a foundation for a conceptual framework for geographic data integration. In GIS-applications (as well in other non-GIS-applications) the crucial characteristic of a piece of information is what it is about, the entities it refers to. It is this referential meaning that needs to be made explicit and organized (Guarino 1997). The key issue in geographic data set integration is finding corresponding instances. This process of semantic matching is only possible if the meaning of objects is clear. Central in a conceptual framework for integration is a mechanism that makes object definitions clear; that means, make data sets semantically transparent to each other. In that respect geographic data set integration can be seen as a communication problem. Any successful communication requires a language that builds on a core of shared concepts (Kuhn 1996). It is here that an ontology plays a fundamental role.

Concept and Definition of an Ontology

The notion and use of an ontology is relatively young, although the term 'ontology' has a long history in philosophical tradition in conceiving ontology as the science, which deals with the nature and organization of reality (Smith 1996).[26] However, in Artificial Intelligence (AI), a subfield of computer science, an ontology has to do with the explication of knowledge to overcome the problem of semantic diversity of different information sources (Wache et al 2001; Pundt and Bishr 2002). In this research the definition of an ontology is made operational as 'a structured, limitative collection of unambiguously defined concepts' (Mars 1995; van der Vet and Mars 1998).

An ontology for a certain discipline is called a domain ontology. Geographic data sets studied here are from the discipline of topographic mapping. In a domain ontology for topographic mapping, definitions for topographic concepts are supplied, such as 'road', 'railway', or 'building'.

An ontology for a certain geographic data set is called here an application ontology. In geographic data sets, names or labels for mapped or surveyed concepts are used, such as 'road' or 'building', but their precise meaning is not necessarily the same as similar names for concepts in the domain ontology. That's why we must make a clear distinction between concepts in the domain ontology, and concepts in application ontologies for the data sets involved in the integration process. This distinction also resolves naming diversity, like homonyms (same name used for different concepts), or synonyms (different names used for same concept).

Surveying Rules

Abstracting the real world is a two-step process.

There exist classes of real-world phenomena. There may be many classes of real-world phenomena, or terrain objects, but only terrain objects from classes, relevant for a certain discipline, which can be identified and labelled, are included

---

[26] Ontology is a greek word. The founding father of the doctrine of existence was the greek philosopher Parmenides. The term ontology was coined by Clauberg in 1646 to indicate the traditional philosophy of Aristotle in Metafysica, one of Aristotle's major works.

as concepts, or classes, in a domain ontology.[27] Rules which govern this selection — from classes of terrain objects into classes of the domain ontology — are defined as abstraction rules.

With this collection of classes we look at the terrain: it is as if we wear a pair of glasses, where only instances of classes of the domain ontology are passed through. From this filtered collection of terrain objects, only those relevant for a certain application are included in an application ontology, and acquired or 'captured' into a geographic data set. Surveying rules (or, alternatively acquisition rules) are defined as rules, which govern the transformation process from the actual observed terrain objects, defined as instances from classes in the domain ontology, into instances of geographic data set classes, as defined in an application ontology.

A Reference Model

In order to integrate different geographic data sets, a domain ontology is refined and restructured into a reference model. The refinement of classes for a reference model depends on classes in application ontologies. The approach here is to include in the reference model information from surveying rules to the level of data classes (Molenaar 1998). Data classes are created by making discrete the value of an attribute by choosing useful limits. For example, domain class 'road' is refined into three data classes: roads with (a) tracks ≤ 2 meters wide, (b) tracks 2 to 4 meters wide, and (c) tracks > 4 meters wide.

Reference Models and Semantic Relationships

Relationships between reference model classes, and application ontology classes, define the semantics of a geographic data set. With these relationships, we define relationships between classes from different application ontologies. These are based on two abstraction mechanisms:

- there is a generalization/specialization classification, which means that reference model classes are grouped into a taxonomy with superclasses and subclasses.
- there is a composite/component classification, which means that reference model classes are grouped into a partonomy, with composite and component classes.

Semantic Similarity

Three types of semantic similarity between classes from different data sets are distinguished:

---

[27] To be as general as possible we use the term class as synonymous of concept.

- equivalent classes: classes from different sets referring to the same reference model class.
- classes with a 'subclass-superclass' relationship: classes from different sets referring to subclass-superclass structure in the reference model.
- classes with a 'composite class-component class' relationship: classes from different sets referring to a composite class-component class structure in the reference model.

Concepts introduced so far — domain ontology, application ontology, abstraction rules, surveying rules, reference model, and semantic relationships — are now configured into a conceptual framework for ontology-based geographic data set integration: Upper-left and upper-right in Figure 24 are geographic data sets to be integrated ('Data Set 1' and 'Data Set 2'). Both data sets have their populations ('Instances Data Set 1' and 'Instances Data Set 2') and their concepts ('Concepts Data Set 1' and 'Concepts Data Set 2'), which are defined, and documented in application ontologies ('Application ontology DS1' and 'Application ontology DS2').

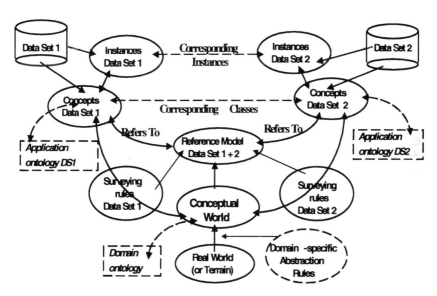

Figure 24: An ontology-based framework for geographic data set integration

Surveying rules capture relevant classes for an application ('Surveying rules Data Set 1' and 'Surveying rules Data Set 2' in Figure 24). Surveying rules are expressed between domain ontology classes, and application ontology classes.

A reference model is constructed based on domain ontology classes, information from surveying rules, and application ontologies classes ('Reference Model Data Set 1 + 2' in Figure 24).

The semantics of data sets is defined by the relationships between reference model classes, and data set classes ('Refers To' in Figure 24).

At the bottom of Figure 24 is the real world (or, terrain). From this terrain, real-world phenomena, of interest, with certain properties, are grouped by abstraction rules in a class, defined as classes in a conceptual world, and documented in a 'domain ontology'.

Constructing a Reference Model

In the previous section the concept of a reference model was explained. The idea of a reference model is to express, or make clear, semantic interconnectedness of data set classes. Basic mechanisms for expressing this semantic interconnectedness are the generalization/specialization classification ('is-a'), and the composite-component classification ('part-whole'). Essentially, a reference model is a subset of concepts from a domain ontology with additional structure. The structure is determined by the concepts of two different application ontologies. Now the construction of a reference model for two geographic data sets is demonstrated.

Geographic Data Sets

The geographic data set integration process is investigated between two geographic data sets, GBKN and TOP10vector:

- GBKN data set is a Dutch large-scale data set (scale 1 : 1,000), a nationwide mapping of buildings, roads, railways and waterways (Table 7).
- TOP10vector data set is a Dutch mid-scale data set (scale 1 : 10,000), a nationwide mapping of buildings, roads, railways, waterways and land use (Table 8).

Domain Ontology Concepts

Comparing GBKN and TOP10vector data sets gives the impression that a small set of concepts will suffice for a domain ontology (Table 10). According to Table 10, as far as GBKN and TOP10vector are concerned, the real world (or terrain) is broken down into six classes. Four of these classes (building, road, water, and land) are refined.

Refining Domain Ontology Concepts with Surveying Rules

Domain ontology concepts from Table 10 are refined into classes for the reference model. This refinement is based on information from surveying rules. In this way we create a common universe of discourse (Table 9). See for details (Uitermark 2001).

A Guiding Principle for Reference Model Construction

After making explicit GBKN and TOP10vector surveying rules and comparing GBKN and TOP10vector data sets, there will be an indication, which classes can be seen as 'subclass-superclass', or 'component class-composite class' to each other, i.e. what role an application class has with respect to another application class. With the domain ontology classes in Table 9 as 'building blocks' we express these roles between application classes in a reference model.

Table 7: GBKN class labels and class descriptions

GBKN class label	Description
Hoofdgebouw	Mainbuilding (building with one or more postal addresses)
Bijgebouw	Annex (building without address)
Rijbaan	Road
Bermsloot	Ditch, less than six meters wide
Spoorbaan	Railway
Inrichtingselement	Verge, flowerbed, parkingstrip, sidewalk
Terrein	Anything but Hoofdgebouw, Bijgebouw, Rijbaan, Bermsloot, Spoorbaan, or Inrichtingselement

Table 8: TOP10vector class labels and class descriptions

TOP10 label	Description	TOP10-label	Description
1000	Mainbuilding or annex	3603	Cycletrack
1050	Barn	4000	Railway
1073	Greenhouse	5023	Wood land
3103	Road, track ¤7m wide	5203	Arable land
3203	Road, track 4-7m wide	5213	Grass land
3303	Road, track 2-4m wide	5263	Anything but 1000, 1050, 1073, 3103, 3203, 3303, 3533, 3603, 4000, 5023,
3533	Street		5203, or 5213

Table 9:   **Domain ontology classes and their refinement into reference model subclasses for GBKN and TOP10vector**

Domain ontology class	Refined subclass in reference model	Definition of refined subclass in reference model	
Building	mainbuilding	**Building** with one or more addresses	
	adjacent annex	**Building** without address connected with 'mainbuilding'	
	free standing annex	**Building** without address not connected with 'mainbuilding'	
		barn	'free standing annex' with a roof on poles with not more than one wall
		greenhouse	'free standing annex' mainly made of glass
		remaining free standing annex	'free standing annex' neither 'barn' nor 'greenhouse'
Road	cycletrack	**Road** for cyclists	
	conngt7m	**Road** ≥ 7 meters wide for local interconnecting traffic	
	conngt4m	**Road** between 4 and 7 meters wide for local inter-connecting traffic	
	conngt2m	**Road** between 2 and 4 meters wide for local inter-connecting traffic	
	street	**Road** in urban area, not for local interconnecting traffic	
Water	ditch	**Water** ≤ 6 meters wide, and interconnecting other **Water**	
Railway		leveled part of the terrain for traffic on rails	
Land	sidewalk	paved strip of **Land** adjacent to **Road** for pedestrians	
	flowerbed	strip of **Land** adjacent or inside 'sidewalk', planted with grass, flowers, or shrubs	
	parkingstrip	paved strip of **Land**, adjacent to **Road** as a provision for parking cars	
	verge	strip of **Land**, on one side adjacent to **Road**	
	arable land	**Land** where agricultural products are cultivated	
	grass land	**Land** mainly overgrown with a grass like vegetation	
	wood land	**Land** overgrown with such a number of leaf wood trees that their crowns form more or less a closed unity	
Otherland		**Land**, not 'sidewalk', 'flowerbed', 'parkingstrip', 'verge', 'arable land', 'grass land', or 'wood land'	

To facilitate the construction of the reference model (its taxonomy subgraph and partonomy subgraph), a guiding principle is presented:

- determine for every application class its role in a semantic similarity. If its role is in a equivalent relationship, then identify its reference model class, and put it in the taxonomy subgraph. If it is in a 'subclass-superclass' relationship, then identify its reference model classes, create a new reference model superclass, and put it in the taxonomy subgraph. If it is in a 'composite class-component class' relationship, then identify its reference model classes, create a new reference model composite class, and put it in the partonomy subgraph.
- determine for every reference model class its relationship with classes in application ontologies. Again, for details see (Uitermark 2001).

Table 10: Six domain ontology concepts and their definition

Class label	Domain ontology concept definition
Building	free standing covered area, partly or completely enclosed by walls, allowing access by people, and directly or indirectly connected to the terrain
Road	leveled part of the terrain for traffic on land
Railway	leveled part of the terrain for traffic on rails
Water	part of the terrain covered by water
Land	part of the terrain, having a distinct use or function, not being building, road, railway, or water
Otherland	**Land**, not having a distinct use or function

Conclusions

This chapter presented a conceptual framework for geographic data set integration. Starting point in this framework is a mechanism to express meaning of geographic data sets in a language of shared concepts, a domain ontology. With references from concepts in data sets to concepts in a domain ontology, semantic matching is accomplished. Concepts of a domain ontology were structured in a reference model to express levels of abstraction between data sets. The approach here is to construct a reference model in such a way that it gives precise information about semantic relationships between classes of different data sets. A consequence of this approach is that all instances will be involved in some correspondence relationship, even with domain classes that are acquired for a single data set. Therefore, if certain instances — so-called singletons — do not take part in a correspondence with other instances then these singletons indicate most probably surveying rule errors.

Now this approach seems linear, but it is not. It is cyclic, and iterative. Even more cyclic, and iterative is the construction of the reference model. The idea is to design a structure that is semantically rich and grained finely enough, to express every semantic similarity between data sets. To facilitate the design of a 'guiding principle', a heuristic was presented. Central in this 'guiding principle' is the concept of role. A role is what a data set class is in confrontation with another data set class: equivalent class, subclass, superclass, component class, or composite class. '

References

Devogele, T., Trevisan, J., and Raynal, L. (1996). *Building a multi-scale database with scale-transition relationships.* Proceedings 7th International symposium on Spatial Data Handling SDH'96 (M.J. Kraak and M. Molenaar, eds.). Delft, The Netherlands, August, 12-16. International Geographical Union. Vol. I, pp. 6.19-6.33.

Guarino, N. (1997). *Semantic matching: formal ontological distinctions for information organization, extraction, and integration.* Proceedings International Summer School, SCIE-97 (M.T. Pazienza, ed.). Frascati, Italy. Lecture Notes in Computer Science, Vol.1299. Springer, Berlin, pp. 139-170.

Kilpeläinen, T. (2000). 'Maintenance of multiple representation databases for topographic data'. *The Cartographic Journal*, Vol. 37, No. 2, pp. 101-107.

Kim, C.-J. (1999). 'Implementation of semantic translation for finding the corresponding geometric objects between topographic databases'. Master Thesis, International Institute for Aerospace Survey and Earth Sciences (ITC), Enschede, The Netherlands.

Kuhn, W. (1996). Semantics of geographic information. Geoinfo-series, Vol.7. Vienna: Department of Geoinformation, Technical University.

Laurini, R. (1998). 'Spatial multi-database topological continuity and indexing: a step towards seamless GIS data interoperability'. *Int. J. Geographical Information Science*, Vol. 12, No. 4, pp. 373-402.

Mars, N.J.I. (1995). *What is an ontology?* Proceedings Seminar on the impact of ontologies on reuse, interoperability and distributed processing (A. Goodall, ed.). London, November 7. Unicom, Uxbridge, Middlesex, UK, pp. 9-19.

Molenaar, M. (1998). An introduction to the theory of spatial object modelling for GIS. London: Taylor and Francis.

Pundt, H. and Bishr, Y. (2002). 'Domain ontologies for data sharing-an example from environmental monitoring using field GIS'. *Computers & Geosciences*, Vol. 28, pp. 95-102.

Sester, M., Anders, K.-H., and Walter, V. (1998). 'Linking objects of different spatial data sets by integration and aggregation'. *GeoInformatica*, Vol. 2, No. 4, pp. 335-357.

Smith, B. (1996). 'Mereotopology: a theory of parts and boundaries'. *Data & Knowledge Engineering*, Vol. 20, No. 3, pp. 287-303.

Uitermark, H.T. (2001). 'Ontology-based geographic data set integration'. PhD Thesis. Computer Science Department, University of Twente, Enschede, The Netherlands.

Uitermark, H.T., van Oosterom, P.J.M., Mars, N.J.I., and Molenaar, M. (1998). *Propagating updates: finding corresponding objects in a multi-source environment.* Proceedings 8th International Symposium on Spatial Data Handling SDH'98 (T.K. Poiker and N. Chrisman, eds.). Vancouver, Canada, July 11-15. International Geographical Union, pp. 580-591.

van der Vet, P.E. and Mars, N.J.I. (1998). 'Bottom-up construction of ontologies'. *IEEE Transactions on Knowledge and Data Engineering*, Vol. 10, No. 4, pp. 513-526.

van Wijngaarden, F.A., van Putten, J.D., van Oosterom, P.J.M., and Uitermark, H.T. (1997). *Map Integration. Update propagation in a multi-source environment.* Proceedings 5th

ACM Workshop on Advances in Geographic Information Systems ACM-GIS'97 (R. Laurini, ed.). Las Vegas, Nevada, USA, November 13-14. ACM, New York, pp. 71-76.

Vogels, A.B.M. (1999). 'Propagatie van GBKN-wegenmutaties naar de TOP10vector (*in Dutch*)'. Master Thesis. Geodesy Department, Technical University Delft, Delft, The Netherlands.

Wache, H., Vögele, T., Visser, U., Stuckenschmidt, H., Schuster, G., Neumann, H., and S. Hübner (2001). 'Ontology-based integration of information. A survey of existing approaches'. Proceedings IJCAI workshop: Ontologies and Information Sharing, Seattle, USA, pp. 108-117.

Glossary

Abstraction is the principle of ignoring those aspects of an object that are not relevant to the current purpose in order to concentrate more fully on those that are.

Action is an executable atomic computation that results in a change in state of the system or the return of the value.

Activity (method) is any group of regulated (proceeding) nonatomic operations, which should be executed in order to accomplish a certain task.

Actor represents a role a user can play with regard to a system, or an entity such as database or another system, which resides outside the system or business being modelled.

Analysis is a process of continual learning about the nuances of problem domain and the system responsibilities.

Application represents manipulation and processing of data in support of user requirements.

Application schema is conceptual schema for data required by one or more applications.

Attribute is some property for which each object in a class has its own value (data).

Class is a formal description of one or a set of objects (instances) with a uniform set of attributes, functions (services) and relationships to other classes, and including a description of how to create new objects in the class.

Conceptual modelling (also information modelling) means the process of abstracting and classifying real word objects of the selected part of reality into the constructs that can be represented like classes in the computer or information system database.

Conceptual formalism is a set of modelling concepts used to describe a conceptual model. The conceptual formalism provides the rules, constraints, inheritance mechanisms, events, activities and other elements that make up a conceptual schema language. Conceptual formalism can be expressed in several conceptual schema languages.

Conceptual model defines concepts of a universe of discourse. Conceptual model provides the abstract description of the selected real-world features.

Conceptual schema is formal description of a conceptual model in conceptual schema language. The conceptual schema is a rigorous description of a conceptual model for some universe of discourse. A conceptual schema language is based upon a conceptual formalism.

Conceptual schema language is formal description technique used, which bases on a conceptual formalism for the purpose of representing conceptual schemas. A conceptual schema language provides the semantic and syntactic elements used to describe the conceptual model rigorously in order to convey meaning consistently. A conceptual schema language may be lexical or graphical or both.

Conceptualization is an abstract and simplified view of the world that we wish to represent for a certain purpose, and which consists of object types and relationships among them that are assumed to exist in some area of interest.

Data model represents the abstract notion and conceptual interpretation of facts or knowledge for the portion of the complex real world and regarding the particular application in mind.

Dataset is identifiable collection of data.

Domain is real, abstract or hypothetical field of endeavour under consideration, which can include various groups of objects that behave according to the rules and characteristics of this domain.

Dynamic model describes the system regarding its subsistence, stability and variations trough time.

Epistemology is the science about knowledge and knowing.

Feature is an abstraction of real world phenomena. A feature may occur as a type (class) or an instance (object).

Feature attribute is an important characteristic of a feature. A feature attribute has a name, a data type, and a value domain associated to it.

Feature catalogue (objects catalogue) is a catalogue containing definitions and descriptions of the feature types (classes), feature attributes, and feature relationships occurring in one or more sets of geographic data, together with any feature operations that may be applied.

Feature operation (method) is an activity that every instance (object) of a feature type (class) may perform.

Function is a synonym for service and is a constituting part of a process. Function consists of an activity or a set of activities, which that process must perform.

Geographic information service is an activity that transforms, manages, or presents geographic data to users.

Geographic information system (GIS) is the combination of skilled persons, spatial and descriptive data, analytical methods, computer software and hardware that are organized to manage and automatically process data with the aim to deliver information to users through the geographic data presentation.

Information system is a combination of database, human and technical resources that together with the appropriate organization skills produce information needed for its users, in order to support certain economic activity, management of resources and/or decision-making procedures.

Land cadastre is a public and systematic registration of land parcels for a certain administrative unit (cadastral jurisdiction). The basic role of land cadastre is identification of real estate properties in space. Therefore it forms the basic technical support for the operation of the land registry. Delineation of parcels is based on measured boundary points and lines.

Land registry is a public database of titles on properties that are divided on real rights (ownership and usage), encumbrances (easements and mortgage) and obligatory rights (leasehold, tenancy, rent, redemption etc.), their changes and legal status.

Message is a selection of one of the class functions that an object knows how to perform.

Metadata are data about data or their interpretation, which describes their technical and administrative characteristics.

Metadata schema is a conceptual schema describing metadata.

Metamodel is model's model that serves for explanation and definition of relationships among the various components of the applied model itself.

Methodology is a set of rules, methods and practical procedures that are used in the specific science or technological discipline. Methodology is science about methods.

Model is an applied abstract supplement, which is formed of descriptive and/or graphical specification of the selected part of reality, and represents simplified mapping of physical environment into the conceived and interpreted notion. Models help to understand and shape both the problem and its solution domain.

Nominal ground is the view of the real world implied by the specification of the geographic dataset. The nominal ground forms the ideal geographic dataset to which the actual geographic dataset is compared regarding to their location, thematic and temporal attributes.

Object is anything in a problem domain, real or abstract, reflecting the capabilities of a system to keep data about it and interacts with it. Each object is an instance of particular abstract object type or class.

Object type is a collection of objects, which can be described with the same attributes (data), services (processing behavior) and relationships. The realization of object type in a particular setting is a class.

Ontology is an explicit specification of conceptualization. Ontology in philosophy refers to the subject of existence (or being). Ontology is a description of the concepts and relationships that can exist for an actor or a community of actors. Ontology is normally given as a set of definitions of formal vocabulary.

Parcel is a set of lots that are all the subject matter of a unit of real estate.

Problem domain is the considered area that encompasses real world features and concepts related to the problem, which the system is being designed to solve.

Process is a flow of development in which something is forming, acting or changing. Process can be regarded as an application of a method in a certain industrial activity that is going on in the real environment.

Quality schema is a conceptual schema defining aspects of quality for geographic (spatial) data.

Reality is considered as an infinite space and time, which we conceive as the complex physical actuality that surrounds us and constantly changes.

Responsibility is the condition, quality, fact or instance of being accountable, answerable or liable, as for a person, trust, service, office or debt.

Schema is a formal description of a model.

Service is a synonym for function and is a specific processing activity or a set of activities that an object is responsible for exhibiting.

State is a condition or situation during the life of an object or performance of an information system, during which it satisfies some condition, performs some activity or waits for some event.

Structure is the manner of organization, and is an expression of problem domain complexity, which is pertinent to the system's responsibilities.

System is structured arrangement of elements or mechanisms that are related or connected as to form unity, in order to achieve their efficient functionality.

System analysis is a combined process dissecting the system responsibilities that are based on the problem domain characteristics and users' requirements.

System requirements are an arrangement of things accountable and related together as a whole that represent all the responsibilities the system must manifest and fulfil.

Universe of discourse is a view of real or hypothetical world that includes everything of interest. A universe of discourse consists of a set of selected object types and the relevant relationships among them. A universe of discourse is described in a conceptual model and formally specified in the conceptual schemas.

Use case is a sequence of actions that an actor (a person, external entity or another system) performs within a system to achieve a particular goal. Or from inside out, a use case is a sequence of actions a system executes that yield observable results of value to a particular actor. A use case specification contains the main, alternate and the exception flows or paths.

Index

Printed and bound by CPI Group (UK) Ltd, Croydon, CR0 4YY

21/10/2024

01777088-0013